INDOOR
AIR POLLUTION

INDOOR
AIR POLLUTION

CHARACTERIZATION, PREDICTION, AND CONTROL

RICHARD A. WADDEN

and

PETER A. SCHEFF

Environmental and Occupational Health Sciences
School of Public Health
University of Illinois
Chicago

A Wiley-Interscience Publication
JOHN WILEY & SONS
New York   Chichester   Brisbane   Toronto   Singapore

Library of Congress Cataloging in Publication Data:

Wadden, R. A.
   Indoor air pollution.

   (Environmental science and technology, ISSN 0194-0287)
   "A Wiley-Interscience publication."
   Includes index.
   1. Air–Pollution, Indoor.   I. Scheff, Peter A.
II. Title.   III. Series.
TD883.1.W33   1982   363.7′392   82-11153
ISBN 0-471-87673-9

SERIES PREFACE

Environmental Science and Technology

The Environmental Science and Technology Series of Monographs, Textbooks, and Advances is devoted to the study of the quality of the environment and to the technology of its conservation. Environmental science therefore relates to the chemical, physical, and biological changes in the environment through contamination or modification, to the physical nature and biological behavior of air, water, soil, food, and waste as they are affected by man's agricultural, industrial, and social activities, and to the application of science and technology to the control and improvement of environmental quality.

The deterioration of environmental quality, which began when man first collected into villages and utilized fire, has existed as a serious problem under the ever-increasing impacts of exponentially increasing population and of industrializing society. Environmental contamination of air, water, soil, and food has become a threat to the continued existence of many plant and animal communities of the ecosystem and may ultimately threaten the very survival of the human race.

It seems clear that if we are to preserve for future generations some semblance of the biological order of the world of the past and hope to improve on the deteriorating standards of urban public health, environmental science and technology must quickly come to play a dominant role in designing our social and industrial structure for tomorrow. Scientifically rigorous criteria of environmental quality must be developed. Based in part on these criteria, realistic standards must be established and our technological progress must be tailored to meet them. It is obvious that civilization will continue to require increasing amounts of fuel, transportation, industrial chemicals, fertilizers, pesticides, and countless other products; and that it will continue to produce waste products of all descriptions. What is urgently needed is a total systems approach to modern civilization through which the pooled talents of scientists and engineers, in cooperation with social scientists and the medical profession, can be focused on the development of order and equilibrium in the presently disparate segments of the human environment. Most of the skills and tools that are needed are already in existence. We surely have a right to hope a technology that has created such manifold environmental problems is also capable of solving them. It is our hope that this Series in Environmental Sciences and Technology will not only serve to make this challenge more explicit to the established professionals, but that it also will help to stimulate the student toward the career opportunities in this vital area.

Robert L. Metcalf
Werner Stumm

PREFACE

Recognition of indoor pollution probably started when the first fire was lit in an unventilated cave. Later, as the enclosure and control of combustion became more sophisticated, most indoor pollutants were vented outside thereby making indoor spaces, if not the ambient environment, more or less habitable. In our day adequate ventilation is taken for granted in industrialized societies. But, with the advent of increasing costs and less plentiful supplies of energy, more emphasis has been placed on using energy efficiently. "Tight" housing, recirculated air, and increased application of insulation are some of the measures promoted to achieve this goal. And predictably, if perhaps unexpectedly, such approaches have sometimes produced pollution concentrations in internal spaces which are hazardous to human health and depressing to the quality of life.

The purpose of this book is to review the information we now have about indoor air pollution hazards and to supply those responsible for design and evaluation with tools for characterizing the problem and its solution. Our goal is to attempt to contribute, as many others have already done, to a heightened awareness of what considerations need to be taken into account. The thrust of our effort has been to describe the indoor air environment of domestic and public buildings, although the techniques are applicable to any indoor space.

The book is organized into four parts: Characterization, Prediction, Control, and Application. After the introductory chapter, Chapters 2 to 5 explore health implications, external and internal contributions, and measurement of indoor air pollution. Chapter 6 outlines the present status of prediction techniques, while Chapter 7 is a summary of the most common control methods. Chapters 8 to 11 explore, in some detail, the application of modeling techniques to several typical indoor settings. The intent of these calculations is to give the user some appreciation of how the background information can be integrated into a realistic evaluation.

As with any rapidly changing field, "last words" are unrealistic and continual updating is a necessary task. However, we believe that the general approach, which focuses on the pollutant mass balance for internal space, will continue to have considerable utility.

The core of this book was developed for a course we offered under the aus
pices of the Air Pollution Control Association. We appreciate the Associatior
giving us a forum to present our ideas. We also thank the secretarial staff of the
Environmental and Occupational Health Sciences program area, and particularly
Lucille Vaughn, for their patient typing of numerous manuscript drafts. And.
finally, we would like to acknowledge the many nonformal contributions made
by the faculty and students of the University of Illinois, School of Public Health.

RICHARD A. WADDEN
PETER A. SCHEFF

Chicago, Illinois
October 1982

CONTENTS

INDOOR
AIR POLLUTION

1

DESCRIBING THE PROBLEM

The quality of the air we breath and the attendant consequences for human health are influenced by a variety of factors. These include hazardous material discharges indoors and outdoors, meteorological and ventilation conditions, and pollutant decay and removal processes. Over 80% of our time is spent in indoor environments (Dockery and Spengler, 1981a; Szalai, 1972; NAS, 1981) so that the influence of building structures, surfaces, and ventilation are important considerations when evaluating air pollution exposures.

Recognition that the indoor air environment is not an exact reflection of outdoor conditions is of relatively recent emergence. The impact of cigarette smoking, stove and oven operation, and emanations from certain types of particleboard, cement, and other building materials are often the most significant determinants of indoor air quality. There is a continuing need to characterize human exposures both from the standpoint of meeting pertinent ambient and occupational standards and the recognition of potential hazardous levels of pollutants which do not have applicable standards. The implications of indoor air concentrations for epidemiological studies, where exposures are based on outdoor measurements, have been recognized and, in the recent past, partially investigated (Spengler et al., 1979; Dockery and Spengler, 1981a, b).

The impact of energy conservation on inside environments may be substantial, particularly with respect to decreases in ventilation rates (Hollowell et al., 1979a) and "tight" buildings constructed to minimize infiltration of outdoor air (Woods, 1980; Hollowell et al., 1979b). Potential conflicts between air quality and energy economics are apparent in existing building codes and efforts to resolve these are continuing (Woods et al., 1981). There is a need for proper building design, construction, and ventilation guidelines to avoid the exposure of inhabitants to unhealthy environments.

Recently a variety of indoor-air measurement studies have been carried out (e.g., Yocum et al., 1977; Yocum, 1982; Dockey and Spengler, 1981a; Hollowell et al., 1979b; Moschandreas et al., 1978). These have done much to better characterize the types and levels of indoor pollutants that occur in the urban settings

1

Table 1.1 Typical Indoor Pollutant Concentrations

Pollutants of Concern	Concentration (Sampling Time)	Location	Reference
Carbon monoxide, CO	2.5–28 ppm	Offices, restaurants, bars, arenas	Table 2.6
	3.1–7.8 ppm (seasonal averages of 12-h samples)	Kitchen of homes with gas stoves	Wade et al. (1975)
Nitrogen dioxide, NO_2	0.005–0.317 ppm (1 wk)	English homes with gas cookers	Table 2.5 Goldstein et al. (1979) Florey et al. (1979)
	0.005–0.11 ppm (24 hr)	American homes with gas stoves	Table 2.5
	<0.06 ppm (24 hr)	American homes with electric stoves	Keller et al. (1979a, 1979b)
Respirable particles, RP	100–700 $\mu g/m^3$ (8–50 min) 20–60 $\mu g/m^3$ (1–42 min)	Restaurants, sport arenas, residences with smoking without smoking	Table 2.6 Repace and Lowrey (1980)
	10–70 $\mu g/m^3$ (24 hr)	Residences	Figure 6.12 Dockery and Spengler (1981a) Spengler et al. (1981)
Total suspended particles, TSP	39–66 $\mu g/m^3$ (averages of 12-hr samples; 26–72% of outdoor concentrations)	Homes, public buildings	Yocum et al. (1977)
	2.7–79.4 $\mu g/m^3$ (48 hr)	Urban hospital	Neal et al. (1978)
Asbestos	0–100 ng/m^3 (0–2 × 10^4 fibers/m^3) (5 min to 10 hr)	Normal activities	Table 4.11 Sawyer and Spooner (1978)
	20 × 10^6 fibers/m^3	During maintenance	Sawyer (1977)

Substance	Concentration	Location	Reference
Formaldehyde, HCHO	60–1673 ppb (~1 hr; 463 ppb average for all measurements)	Homes with chipboard walls	Andersen et al. (1975)
	30–1770 ppb (35–60 min)	Mobile homes	Breysse (1981)
Ozone, O_3	<0.002–0.068 ppm (40 min to 2 hr)	Photocopying machine room	Selway et al. (1980)
	<0.002–0.018 ppm (30 min)	Homes with electrostatic air cleaners	Allen et al. (1978)
Radon, Ra-222	0.005–0.94 pCi/L (0.01 pCi/L average)	House in Boston	UN (1977)
	≤ 25–34 pCi/L (averages of 3- to 6-min samples)	House on Florida reclaimed phosphate land	Windham et al. (1978)
Radon daughters	0.003–0.013 WL[a] (average: 0.004 WL)	Houses in New York, New Jersey	NAS (1981)
	0.005–0.05 WL (average: 0.01 WL) (averages of >24-hr to 1-wk samples)	Houses on reclaimed phosphate land in Florida	Guimond et al. (1979)
Benzo(a)pyrene	7.1–21.0 ng/m³ (~2–4 hr)	Sports arena	Table 2.6 Elliott and Rowe (1975)
Dimethylnitrosamine	0.11–0.24 µg/m³ (90 min)	Bar	Table 2.6 Brunneman and Hoffmann (1978)
Carbon dioxide, CO_2	0.086% (5 min)	Lecture hall	Wang (1975)
	0.06–0.25%	School room	Kusuda (1976)
	0.9% (continuous measurements for ~8 wk)	Nuclear submarines	Wilson and Schaeffer (1979)
Viable particles	20–700 CFP/m³ (averages of 10-min samples taken every 40 min)	Schools, hospitals, residences	Berk et al. (1980)

[a]WL, working level.

3

of industrialized countries. Table 1.1 indicates typical nonindustrial indoor concentrations for some pollutants of concern.

While these data supply useful points of reference, it is often difficult to extrapolate such measurements to other conditions of ventilation, outdoor air pollution, meteorology, and indoor sources. In order to overcome some of these limitations we have used a pollutant mass balance model for an interior space as a focus for our discussion. The general form of such a balance is

$$\begin{pmatrix} \text{pollutant} \\ \text{flow in} \end{pmatrix} - \begin{pmatrix} \text{pollutant} \\ \text{flow out} \end{pmatrix} + \begin{pmatrix} \text{source} \\ \text{emissions} \end{pmatrix} - \begin{pmatrix} \text{sink} \\ \text{removals} \end{pmatrix} = \begin{pmatrix} \text{indoor pollutant} \\ \text{accumulation} \end{pmatrix}$$

(1.1)

each of the contributions having the dimensions of mass/time. The pollutant flow terms may be due to outdoor air infiltration or forced convection, recirculation, or a combination of these. If appropriate values are available for each of the factors in Equation (1.1), the indoor concentration can be estimated for a broad spectrum of conditions. The use of this tool then makes possible the

Table 1.2   Air Quality Standards Promulgated by the United States Environmental Protection Agency[a]

	Standard Concentration		
	$\mu g/m^3$	ppm	Averaging Time
Suspended particulate matter, TSP	75 (60)[b]	–	Annual geometric mean
	260 (150)[b]	–	24-h
Sulfur dioxide, SO_2	80	0.03	Annual mean
	365	0.14	24-h
	(1300)[b]	(0.50)[b]	3-h
Carbon monoxide, CO	10,000	9	8-h no more than once per year
	40,000	35	1-h no more than once per year
Nitrogen oxides, NO_2	100	0.05	Annual mean
Ozone, O_3	235	0.12	1-hr daily maximum no more than once per year
Nonmethane hydrocarbons	160	0.24	6–9 a.m. average no more than once per year
Lead, Pb	1.5	–	3-month average

[a]EPA (1971, 1978, 1979).
[b]Secondary standards in parentheses.

comparison of a variety of alternatives for control of an indoor pollution problem.

In order to protect the public health, air quality standards for outdoor air have been promulgated in the United States for total suspended particulate matter, carbon monoxide, nitrogen oxide, sulfur dioxide, ozone, nonmethane hydrocarbons, and lead (EPA, 1971, 1978, 1979) (Table 1.2). Indoor air quality has been specified for industrial environments for a variety of hazardous substances based on 8- to 10-h days and a 40-h week (ACGIH, 1980). A partial list of occupational standards is given in Table 1.3. In the United States there are presently no legally enforceable health-related national standards for living and recreational spaces or transportation modes. In addition, there is a lack of defini-

Table 1.3   American Conference of Governmental Industrial Hygienists Recommended Occupational Standards[a]

Component	Threshold Limit Values—TWA[b]	
	ppm	mg/m^3
Acetaldehyde	100	180
Acrolein	0.1	0.25
Asbestos	carcinogen	
Amosite	0.5 fiber $>$ 5 μm in length/cm^3	
Chrysotile	2 fiber $>$ 5 μm in length/cm^3	
Crocidolite	0.2 fiber $>$ 5 μm in length/cm^3	
Tremolite	0.5 fiber $>$ 5 μm in length/cm^3	
Other forms	2 fiber $>$ 5 μm in length/cm^3	
Benzene	10	30
Benzo(a)pyrene	Carcinogen—no TWA presently assigned	
Carbon dioxide	5000	9000
Carbon monoxide	50	55
Ethylene glycol methyl ether acetate (methyl Cellosolve® acetate)	25	120
Formaldehyde	2	3
Lead	–	0.15
Methyl chloride	50	105
Mineral dusts	–	5 (respirable dust)
Nitric oxide	25	30
Nitrogen dioxide	3	6
Ozone	0.1	0.2
Sulfur dioxide	2	5

[a]ACGIH (1980).
[b]TWA—time weighted average for normal 8-h day or 40-hr work week.

tion about the governmental agencies responsible for air quality in these areas (Comptroller General, 1980). In general, the Scandinavian countries, faced with cold climates and expensive energy, have been in the forefront of research and legislation pertaining to the indoor pollution implications of energy conservation (e.g., NAS, 1981). But the problems are common to all cultures with extensive energy requirements.

While the characterization and control concepts presented here are applicable to any enclosed volume, the emphasis of our discussion is on the nonindustrial indoor environment. Most of our examples have been drawn from the North American experience. But the elements of evaluation are applicable to other settings as well, with the recognition that cultural differences in life-style will significantly affect indoor pollutant generation, dispersal, and control. For instance, the use of kerosene space heaters in Japanese residences is much more common than central heating systems. Consequently, techniques to predict indoor concentrations need to take into account this additional pollution source and the absence of forced-air ventilation. In addition, the willingness of the Japanese citizen to endure lower indoor temperatures than his Western counterpart, the limited amount of insulation ordinarily available in domestic construction, and the planning of most residences to take advantage of passive heating due to southern exposures are all factors in the Japanese setting which will modify the evaluation.

With the continued increase in energy cost, both present and future dwellings, as well as public and office buildings, will be designed or altered to conserve heat and refrigeration. And these changes call for the application of both existing and novel characterization and control techniques which have been adapted to and evaluated for the nonindustrial setting. It is our hope that this book supplies a basis from which to develop such techniques and that it will provide some useful approaches to understanding, characterizing, and controlling the indoor environment.

REFERENCES

ACGIH (1980). *TLV's threshold limit values for chemical substances in workroom air adopted by ACGIH for 1980*. American Conference of Governmental Industrial Hygienists, Cincinnati, Ohio.

Allen, R. J., Wadden, R. A., and Ross, E. D. (1978). Characterization of potential indoor sources of ozone. *Am. Ind. Hyg. Assoc. J.* 39:466–471.

Andersen, I., Lundqvist, G. R., and Molhave, L. (1975). Indoor air pollution due to chipboard used as a construction material. *Atmos. Environ.* 9: 1121–1127.

Brunneman, K. D. and Hoffman, D. (1978). Chemical studies on tobacco smoke.

LIX. Analysis of volatile nitrosamines in tobacco smoke and polluted indoor environments. In: *Environmental Aspects of N-Nitroso Compounds,* F. A. Walker, M. Castegnaro, L. Griciute, and R. E. Lyle, Eds., IARC Scientific Publication No. 19, Lyon, pp. 343–356.

Berk, J. V., Boyan, T. A., Brown, S. R., Ko, I., Koonce, J. W., Loo, B. W., Pepper, J. H., Robb, A. W., Strong, P. C., Turiel, I. and Young, R. A. (1980). Field monitoring of indoor air quality. In: *1979 Annual Report of the Energy and Environment Division,* Lawrence Berkeley Laboratory, University of California, Report No. LBL 11650.

Breysse, P. A. (1981). The health cost of tight homes. *J. Am. Med. Assoc.* 245: 267–268.

Comptroller General of the United States (1980). *Indoor air pollution: An emerging health problem.* Report to the Congress of the United States, No. CED-80-111, U.S. Government Printing Office, Washington, D.C.

Dockery, D. W. and Spengler, J. D. (1981a). Personal exposure to respirable particulates and sulfates. *J. Air Pollut. Control Assoc.* 31 : 153–159.

Dockery, D. W. and Spengler, J. D. (1981b). Indoor–outdoor relationships of respirable sulfates and particles. *Atmos. Environ.* 15 : 335–343.

Elliott, L. P. and Rowe, D. R. (1975). Air quality during public gatherings. *J. Air Pollut. Control Assoc.* 25 : 635–636.

EPA (1971). Primary and secondary air quality standards. *Fed. Reg.* 36 : 22388–22392.

EPA (1978). National air quality standard for lead. *Fed. Reg.* 43 : 46245–46277.

EPA (1979). Revisions to the national ambient air quality standard for photochemical oxidants. *Fed. Reg.* 44 : 8201–8233.

Florey, C. duV., Melia, R. J. W., Chinn, S., Goldstein, B. D., Brooks, A. G. F., John, H. H., Craighead, I. B., and Webster, X. (1979). The relation between respiratory illness in primary school children and the use of gas for cooking. III—Nitrogen dioxide, respiratory illness and lung infection. *Int. J. Epid.* 8 : 347–353.

Goldstein, B. D., Melia, R. J. W., Chinn, S., Florey, C. duV., Clark, D., and John, H. H. (1979). The relation between respiratory illness in primary school children and the use of gas for cooking. II—Factors affecting nitrogen dioxide levels in the home. *Int. J. Epid.* 8 : 339–346.

Guimond, R. J., Ellett, W. H., Fitzgerald, J. E., Windham, S. T., and Cuny, P. A. (1979). *Indoor radiation exposure due to Radium-226 in Florida phosphate lands.* U.S. Environmental Protection Agency Report No. EPA-520/4-78-013.

Hollowell, C. D., Berk, J. V., and Traynor, G. W. (1979a). Impact of reduced infiltration and ventilation on indoor air quality. *ASHRAE J.,* pp. 49–53, July.

Hollowell, C. D., Berk, J. V., Lin, C., Nazaroff, W. W., and Traynor, G. W. (1979b). Impact of energy conservation in buildings on health. In: *Chang-*

ing Energy Use Futures, R. A. Fazzolare and C. B. Smith, Eds., Pergamon, New York.

Keller, M. D., Lanese, R. R., Mitchell, R. I., and Cote, R. W. (1979a). Respiratory illness in households using gas and electric cooking. I. Survey of incidence. *Environ. Res.* **19**:495–503.

Keller, M. D., Lanese, R. R., Mitchell, R. I., and Cote, R. W. (1979b). Respiratory illnesses in households using gas and electric cooking. II. Symptoms and objective findings. *Environ. Res.* **19**:504–515.

Kusuda, T. (1976). Control of ventilation to conserve energy while maintaining acceptable indoor air quality. *ASHRAE Trans.* **82**(Part 1):1169–1181.

Moschandreas, D. J., Stark, S. W. C., McFadden, J. F., and Morse, S. S. (1978). *Indoor air pollution in the residential environment*, Vols. I and II. U.S. Environmental Protection Agency Report No. EPA-600/7-78-229a&b.

NAS (1981). *Indoor Pollutants*, National Academy of Sciences, Washington, D.C.

Neal, A. deW., Wadden, R. A., and Rosenberg, S. H. (1978). Evaluation of indoor particulate concentrations for an urban hospital. *Am. Ind. Hyg. Assoc. J.* **39**:578–582.

Repace, J. L. and Lowrey, A. H. (1980). Indoor air pollution, tobacco smoke, and public health. *Science* **208**:464–472 (May 2).

Sawyer, R. N. (1977). Asbestos exposure in a Yale building. *Environ. Res.* **13**:146–169.

Sawyer, R. N. and Spooner, C. N. (1978). *Sprayed asbestos-containing materials: A guidance document.* U.S. Environmental Protection Agency Report No. EPA-450/2-78-014.

Selway, M. D., Allen, R. J., and Wadden, R. A. (1980). Ozone production from photocopying machines. *Am. Ind. Hyg. Assoc. J.* **41**:455–459.

Spengler, J. D., Ferris, B. G., Dockery, D. W., and Speizer, F. E. (1979). Sulfur dioxide and nitrogen dioxide levels inside and outside homes and the implications on health effects research. *Environ. Sci. Technol.* **13**:1276–1280.

Spengler, J. D., Dockery, D. W., Turner, W. A., Wolfson, J. M., and Ferris, B. G. (1981). Long-term measurements of respirable sulfates and particles inside and outside homes. *Atmos. Environ.* **15**:23–30.

Szalai, A. (Ed.) (1972). *The Use of Time. Daily Activities of Urban and Suburban Populations in 12 Countries*, Mouton and Co., The Hague.

UN (1977). *Sources and effects of ionizing radiation.* United Nations Scientific Committee on the Effects of Atomic Radiation 1977 Report to the General Assembly, with Annexes, No. E77.1X.1, New York.

Wade, W. A., Cote, W. A., and Yocum, J. E. (1975). A study of indoor air quality. *J. Air Pollut. Control Assoc.* **25**:933–939.

Wang, T. C. (1975). A study of bioeffluents in a college classroom. *ASHRAE Trans.* **81**(Part 1):32–44.

Wilson, A. J. and Schaeffer, K. E. (1979). The effect of prolonged exposure to elevated carbon monoxide and carbon dioxide levels on red blood cell parameters during submarine patrols. *Undersea Biomedical Research* 6(Suppl.): S-49 to S-56.

Windham, S. T., Savage, E. D., and Phillips, C. R. (1978). *The effects of house ventilation systems on indoor radon–radon daughter levels.* U.S. Environmental Protection Agency Report No. EPA-520/5-77-011.

Woods, J. E. (1980). Environmental implications of conservation and solar space heating. Engineering Research Institute, Iowa State University, Ames, Iowa, BEUL 80-3. Meeting of the New York Academy of Sciences, New York, January 16.

Woods, J. E., Maldonado, E. A. B., and Reynolds, G. L. (1981). Safe and energy efficient control strategies for indoor air quality. Engineering Research Institute, Iowa State University, Ames, Iowa, Preprint BEUL 81-01. Meeting of the American Association for the Advancement of Sciences, Toronto, Canada, January 3–8.

Yocum, J. E. (1982). Indoor-outdoor air quality relationships. A critical review. *J. Air. Pollut. Control Assoc.* **32**:500–520.

Yocum, J. E., Coté, W. A., and Benson, F. B. (1977). Effects of indoor air quality. In: *Air Pollution*, Vol. II, 3rd ed., A. E. Stern, Ed., Academic Press, New York.

PART ONE

CHARACTERIZATION OF INDOOR AIR QUALITY

2

HEALTH EFFECTS CRITERIA

Although it is not meant to be a definitive list, the pollutants we will discuss in some detail have been recognized, and to a certain extent characterized, as potential indoor air pollution problems. These include carbon monoxide (CO), nitrogen oxides (NO_x), tobacco smoke components, suspended particulate matter [both total suspended particulate matter (TSP) and respirable particles (RP)], asbestos, formaldehyde (HCHO), ozone (O_3), and radon (Ra-222). Other materials such as carbon dioxide (CO_2), organics, viable particulate matter (TVP), and odor-causing chemicals are also of some concern.

CARBON MONOXIDE

Carbon monoxide is a chemical asphyxiant gas. Its affinity for hemoglobin in red blood cells is 200–250 times that of oxygen, which can result in significant reduction in oxygen-carrying capacity. Carboxyhemoglobin concentration (COHb), expressed as % of saturation, is the biochemical measure usually referred to as an indicator of CO uptake.

The relationship between CO exposure and COHb levels can be expressed by a mathematical equation based on theoretical principles (Coburn et al., 1965; NAS, 1977a):

$$\frac{d(CO)}{dt} = \dot{V}_{CO} - \frac{[COHb]}{[O_2Hb]} \frac{\bar{p}c(O_2)}{M} \frac{1}{1/D_L + (P_B - 47)/\dot{V}_A}$$

$$+ \frac{P_1(CO)}{1/D_L + (P_B - 47)/\dot{V}_A} \tag{2.1}$$

where $d(CO)/dt$ = rate of change of carbon monoxide in the body, ml/min

$\dot{V}_{CO}$ = endogenous carbon monoxide production rate, ml/min

[COHb] = concentration of carbon monoxide in the blood, ml/ml

[O_2Hb] = concentration of oxyhemoglobin, ml/ml

$\bar{p}c(O_2)$ = mean pulmonary capillary oxygen pressure, mm Hg

M = Haldane constant, ratio of the affinity of Hb for CO to that for O_2 (220 for pH 7.4)

D_L = diffusion capacity of the lungs for CO, ml/min $\cdot$ mm Hg

P_B = barometric pressure, mm Hg

$\dot{V}_A$ = alveolar ventilation rate, ml/min

$P_1(CO)$ = inspired carbon monoxide partial pressure, mm Hg.

An approximate solution of Equation (2.1) is (Coburn et al., 1965)

$$\frac{\alpha[COHb]_t - \beta \dot{V}_{CO} - P_1(CO)}{\alpha[COHb]_0 - \beta \dot{V}_{CO} - P_1(CO)} = e^{-\alpha t/\beta V_B} \tag{2.2}$$

where t = duration of exposure, min

$\alpha = \bar{p}c(O_2)/M[O_2 Hb]'$

$\beta = 1/D_L + P_L/\dot{V}_A$

V_B = blood volume, ml

$[COHb]_t$ = concentration of COHb in blood at time t, ml/ml

$[COHb]_0$ = initial concentration of COHb in blood, ml/ml

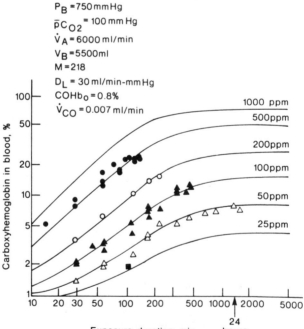

Figure 2.1.   Relationship between exposure duration, ambient carbon monoxide concentration, and carboxyhemoglobin in blood (resting individuals) (Peterson, 1977). Reprinted by permission of Prentice-Hall, Inc. Englewood Cliffs, N.J.

Table 2.1   Levels of Carboxyhemoglobin and Reported Effects[a]

COHb (%)	Effects
0.4	Normal physiologic value for nonsmokers
2.5–3	Decreased exercise performance in patients with angina or with intermittent claudication
4–5	Increased symptoms in traffic policemen (headache, lassitude); increased oxygen debt in nonsmokers
5–10	Changes in myocardial metabolism and possible impairment; statistically significant diminution of visual perception, manual dexterity, or ability to learn
10+	Headache and impaired manual coordination; changes in visual evoked response (VER) by electroencephalogram (EEG)

[a]Ferris (1978).

$$[O_2 Hb]' = (1.38)(g\ Hb/ml\ blood) - [COHb]_t, ml/ml$$
P_L = barometric pressure – partial pressure of water vapor at body temperature, mm Hg

This equation has been successfully tested for human subjects under controlled conditions (Peterson and Stewart, 1975). A typical set of curves based on Equation (2.2) are given for nonsmokers in Figure 2.1.

Tables 2.1 and 2.2 (Ferris, 1978; EPA, 1980) show the progression of effects associated with increasing CO exposures. COHb levels for smokers are typically 4–6% greater than for nonsmokers (EPA, 1979b; NAS, 1977a). Effects on persons with existing cardiovascular diseases, especially those experiencing angina or intermittant claudication, are apparent at relatively low COHb concentrations (Aronow et al., 1974a; EPA, 1979b).

NITROGEN OXIDES

Most health effects associated with nitrogen oxides (NO_x) have been attributed to nitrogen dioxide (NO_2). However, the major anthropogenic sources of NO_x are combustion processes which primarily produce nitric oxide (NO). This gas is subsequently oxidized to NO_2. Levels of NO_2 above 282 mg/m^3 (150 ppm) can be lethal while concentrations in the range of 94–282 mg/m^3 (50-150 ppm) can produce chronic lung disease (Ferris, 1978). The earliest response to NO_2 occurs in the sense organs. Odor can be perceived at 0.23 mg/m^3 (0.12 ppm) and reversible changes in dark adaptation at exposures of 0.14–0.50 mg/m^3 (0.075-0.26 ppm) (NAS, 1977b). Animal studies have suggested that reduced resistance to

Table 2.2 Compilation of Effects Reported in Selected Human Studies Examining Carbon Monoxide Exposures[a]

CO Concentration (ppm)	Exposure Duration (hours)	Percent Carboxyhemoglobin, Group Mean (and Range)	Reported Effect(s)[b]	References
0	—	0.3–0.7	Physiologic norm (endogenous production)	Coburn et al. (1963)
Not reported (passive smoking)	2	1.8 and 2.3	Duration of exercise until onset of angina decreased 22 and 38%	Aronow (1978)
50	2	2.7 (2.5–3.0)	Duration of exercise until onset of angina decreased 16%	Aronow and Isbell (1973)
50	2	2.8 (2.4–3.1)	Duration of exercise until onset of leg pain (intermittent claudication) in persons with peripheral vascular disease decreased 17%	Aronow et al. (1974a)
50	4[c]	2.9 (1.3–3.8)	Duration of exercise until onset of angina decreased 15%	Anderson et al. (1973)
150	d	3.9 (3.4–4.4)	Impairment of ventricular (heart) functioning in anginal patients with coronary heart disease	Aronow et al. (1974b)
100	1	4.0 (3.0–4.9)	Duration of maximal exercise in 10 healthy nonsmokers decreased 5%, with 1 subject exhibiting an electrocardiographic abnormality following CO and exercise	Aronow and Cassidy (1975)
100	1	4.1	Duration of exercise until marked difficulty in breathing (dyspnea) in persons with emphysema decreased 33%	Aronow et al. (1977)
100	4[c]	4.5 (2.8–5.4)	Duration of exercise until onset of angina decreased 15% and duration of pain increased 31%	Anderson et al. (1973)
42–63 (mean 53) 2	1.5[e] 2[f]	5.1 (3.8–8.0) 2.9 (0.6–3.9)	Duration of exercise until onset of angina decreased 30% after freeway exposure and was still decreased 15% 2 hr later	Aronow et al. (1972)

[a] EPA (1980).
[b] Percentages represent the group mean response.
[c] Intermittent.
[d] Not reported.
[e] Freeway.
[f] After return from freeway.

Table 2.3 Effects of Exposure to Nitrogen Dioxide on Pulmonary Function in Controlled Studies of Sensitive Humans[a]

Concentration μg/m³	Concentration ppm	Number of Subjects	Exposure Time	Effects	References
9400	5.0	14 chronic bronchitics	60 min	No change in mean PAO_2,[b] during or after exposure compared with pre-exposure values, but PaO_2[c] decreased significantly in the first 15 min. Continued exposure for 60 min produced no enhancement of effect.	von Nieding et al. (1973)
3800–9400	2.0–5.0	25 chronic bronchitics	10 min	Significant decrease in PaO_2 and increase in $AaDO_2$[d] at 7500 μg/m³ (4.0 ppm) and above; no significant change at 3800 μg/m³ (2.0 ppm).	von Nieding et al. (1971)
940–9400	0.5–5.0	63 chronic bronchitics	30 inhalations	Significant increase in R_{aw}[e] above 3000 μg/m³ (1.6 ppm); no significant effect below 2800 μg/m³ (1.5 ppm).	von Nieding et al. (1971)
940	0.5	10 healthy 7 chronic bronchitics 13 asthmatics	2 hr	One healthy and one bronchitic subject reported slight nasal discharge. Seven asthmatics reported slight discomfort. Bronchitics and asthmatics showed no statistically significant changes for all pulmonary functions tested when analyzed as separate groups but showed small, but statistically significant, changes in quasistatic compliance when analyzed as a single group	Kerr et al. (1979)
190	0.1	20 asthmatics	1 hr	Significant increase in SR_{aw}[f]. Effect of bronchoconstriction enhanced after exposure in 13 of 20 subjects. Neither effect observed in 7 of 20 subjects. A bronchoconstrictor (carbachol) was used.	Orehek et al. (1976)

[a]EPA, 1981.
[b]PAO_2, alveolar partial pressure of oxygen.
[c]PaO_2, arterial partial pressure of oxygen.
[d]$AaDO_2$, difference between alveolar and arterial blood partial pressure of oxygen.
[e]R_{aw}, airway resistance.
[f]SR_{aw}, specific airway resistance.

17

Table 2.4 Effects of Exposure to Nitrogen Dioxide on Pulmonary Function in Community Studies[a]

Measure	NO$_2$ Exposure Concentrations		Study Population	Effect	References
	$\mu g/m^3$	ppm			
Los Angeles: High-exposure group			205 office workers	No differences in most tests. Smokers in both cities showed greater changes in pulmonary function than nonsmokers.	Linn et al. (1976)
Median hourly NO$_2$	130	0.069			
90th percentile NO$_2$	250	0.133			
Median hourly O$_x$[b]		0.15			
90th percentile O$_x$		0.15			
San Francisco: Low-exposure group			439 office workers		
Median hourly NO$_2$	65	0.035			
90th percentile NO$_2$	110	0.058			
Median hourly O$_x$		0.02			
90th percentile O$_x$		0.03			
Boston: High-exposure group			Pulmonary function tests administered to 128 traffic policemen in urban Boston and to 140 patrol officers in nearby suburban areas.	No difference in various pulmonary function tests. Appeared to be a tendency (which was not statistically significant) for men with more years working in traffic to have higher prevalence of respiratory symptoms (Ferris, 1978).	Speizer and Ferris (1973) Burgess et al. (1973)
Mean "annual" 24-hr concentrations[c]:	103+ 92 SO$_2$	0.055+ 0.035 SO$_2$			
One-hour mean:	260–560	0.14–0.30			
Boston: Low-exposure group					
Mean "annual" 24-hr concentrations[c]:	75+ 36 SO$_2$	0.04+ 0.014 SO$_2$			
One hour mean:	110–170	0.06–0.09			

Los Angeles: High-exposure group:		Adult nonsmokers	No differences in several ventilatory measurements including spirometry and flow volume curves.	Cohen et al. (1972)
annual mean 24-hr concentrations	96	0.051		
90th percentile	188	0.10		
Estimated 1-hr maximum[d]	480–960	0.26–0.51		
San Diego: Low-exposure group:		Adult nonsmokers		
annual mean 24-hr concentrations	43	0.02		
90th percentile	113	0.06		
Estimated 1-hr maximum[d]	225–430	0.12–0.23		
Japan: One-hour concentration at time of testing (1:00 p.m.)	40–360	0.02–0.19	20 school children 11 years of age	During warmer part of the year (April–October) NO_2, SO_2 and TSP[e] significantly correlated with V_{max}[f] at 25% and 50% FVC[g] and with specific airway conductance. Temperature was the factor most clearly correlated with weekly variations in specific airway conductance with V_{max} at 25% and 50% FVC. Significant correlation between each of four pollutants (NO_2, NO, SO_2, and TSP) and V_{max} at 25% and 50% FVC; but no clear delineation of pollutant concentrations at which effects occur.	Kagawa and Toyama (1975) Kagawa et al. (1976)

[a]EPA, 1981.
[b]O_x, photochemical oxidants, mainly ozone.
[c]Mean "annual" concentrations derived from 1-hour measurements using Saltzman technique.

[d]Estimated at 5 to 10 times annual mean 24-hour averages.
[e]TSP, total suspended particulate matter.
[f]V_{max}, maximum expiratory flow rate.
[g]FVC, forced vital capacity.

Table 2.5   Effects of Exposure to Nitrogen Dioxide in the Home on the Incidence of Acute Respiratory Disease in Epidemiology Studies Involving Gas Stoves[a]

Pollutant	NO$_2$ Concentration µg/m^3 (ppm)	Study Population	Effects	Reference
		Studies of Children		
NO$_2$ plus other gas combustion products	NO$_2$ concentration not measured at time of study	2554 children from homes using gas to cook compared to 3204 children from homes using electricity. Ages 6–11	Bronchitis, day or night cough, morning cough, cold going to chest, wheeze, and asthma increased in children in homes with gas stoves.	Melia et al. (1977)
NO$_2$ plus other gas combustion products	NO$_2$ concentration not measured in same homes studied	4827 children ages 5–10	Higher incidence of respiratory symptoms and disease associated with gas stoves.	Melia et al. (1979)
NO$_2$ plus other gas combustion products	Kitchens: 9–596 (gas) (0.005–0.317) 11–353 (electric) (0.006–0.188) Bedrooms: 7.5–318 (gas) (0.004–0.169) 6–70 (electric) (0.003–0.037) (by triethanolamine diffusion samplers)	808 6- and 7-year olds	Higher incidence of respiratory illness in gas-stove homes. No apparent statistical relationship between lung function tests and exposure.	Florey et al. (1979) Companion paper to Melia et al. (1979); Goldstein et al. (1979)
NO$_2$ plus other gas combustion products	Sample of households 24-hr average: gas (0.005–0.11); electric (0–0.06): outdoors (0.015–0.05); monitoring location not reported; 24-hr averages by modified Jacobs–Hochheiser (sodium arsenite); peaks by chemiluminescence	128 children 0–5 346 children 6–10 421 children 11–15	No significant difference in reported respiratory illness between homes with gas and electric stoves in children from birth to 12 years. No differences in lung function tests.	Mitchell et al. (1974); See also Keller et al. (1979a, b)

NO$_2$ plus other gas stove combustion products	Sample of same households as reported above but no new monitoring reporting	174 children under 12	No evidence that cooking mode is associated with the incidence of acute respiratory illness.	Keller et al. (1979b)
NO$_2$ plus other gas stove combustion products	95 percentile of 24-hr indoor average: 39–116 μg/m^3 (0.02–0.06) (gas); 17.6–95.2 μg/m^3 (0.01–0.05) (electric); frequent peaks (gas) > 1100 μg/m^3 (0.6 ppm); 24-hr by modified sodium arsenite; peaks by chemiluminescence	8120 children 6–10; 6 different communities; data collected also on history of illness before the age of 2.	Significant association between history of serious respiratory illness before age 2 and use of gas stoves. Small but statistically significant decrements in lung function tests (FEV$_{1.0}$[b] = 16 ml, FVC[c] = 18 ml) for those from gas stove homes compared with children from homes with electric stoves.	Speizer et al. (1980)

Studies of Adults

NO$_2$ plus other gas stove combustion products	Preliminary measurements peak hourly 470–940 μg/m^3	Housewives cooking with gas stoves; compared to those cooking with electric stoves.	No increased respiratory illness	EPA (1976)
NO$_2$ plus other gas stove combustion products	See table above for monitoring	Housewives cooking with gas stoves, compared to those cooking with electric stoves 146 households.	No evidence that cooking with gas associated with an increase in respiratory disease.	Keller et al. (1979a, b)
NO$_2$ plus other gas stove combustion products	See table above for monitoring	Members of 441 households	No significant difference in reported respiratory illness among adults in gas vs. electric cooking homes.	Mitchell et al. (1974); See also Keller et al. (1979a, b)
NO$_2$ plus other gas stove combustion products	See table above for monitoring	Members of 120 households	No significant difference in acute respiratory disease incidence gas vs. electric cooking homes among adults.	Keller et al. (1979a, b)

[a]EPA (1981).
[b]Forced expiratory volume at 1 second.
[c]Forced vital capacity.

respiratory infection is the most sensitive indicator of respiratory damage. Table 2.3 summarizes controlled studies on sensitive humans and Table 2.4 lists community exposure studies. More striking are the data summarized in Table 2.5 which show a small but apparently higher incidence of respiratory symptoms and disease for children living with gas stoves (an NO_x source) versus those in homes with electric stoves. When indoor concentrations were measured, the levels were much lower than were previously thought to contribute to lung function changes or disease effect. These effects were not observed in adults living in the same or similar environments.

TOBACCO SMOKE

Tobacco smoke contains both particulate matter (most of which is in the respirable range <1 μm) and gaseous components. Some of these are listed in Table 2.6 along with typical concentration ranges in experimental chambers and indoor environments. Other components include phenols, naphthalenes, trace metals, hydrogen cyanide, ammonia, and radioactive polonium-210 (HEW, 1979).

While voluntary smoking has been judged harmful to health (HEW, 1979), specific dose–effect relationships between tobacco smoke and health effects have not yet been developed. However, considerable experimental and epidemiological evidence exists which points out the association between involuntary smoking and adverse health conditions. Children living in households where parents smoked have been found to incur adverse pulmonary effects when compared to those from nonsmoking households (Tager et al., 1979). Parental smoking also appears to be a cause of increased respiratory disease in children in the first year of life (Harlap and Davis, 1974; Colley et al., 1974; Leeder et al., 1976). Finnish children (≤5-years old) of smoking mothers had significantly higher morbidity (mostly due to respiratory disease) and were more likely to be hospitalized and for longer times than children of nonsmoking mothers (Rantakallio, 1978a, b).

A study of 90,000 nonsmoking Japanese women indicated that wives of heavy smokers have a higher risk of developing lung cancer than those married to light or nonsmokers (Hirayama, 1981). The same general conclusion was arrived at for a small number of nonsmoking Greek women (Trichopolous et al., 1981). An analysis of 177,000 nonsmoking American women did not reveal statistically significant differences in lung cancer rates between those married to smokers and nonsmokers (Garfinkel, 1981). However, the rates for involuntary smokers were somewhat increased. Cultural conditions may provide significant explanations for the differences between these studies.

Involuntary smoking by patients with coronary heart disease has been shown

Table 2.6 Measurements of Constituents of Tobacco Smoke[a]

Constituent	Location	Ventilation, Air Changes Per Hour	Amount of Tobacco Burned	Concentration	Source[b]
Experimental Conditions					
CO	80–170 m³ rooms	6.4–2.3	46–101 cigarettes	4.5–75 ppm	1, 2
	Small car, 25 m³ chamber	none	4–9	12–110 ppm	3, 4
Nicotine	57–80 m³ rooms	6.4–8.2	42 cigarettes, 9 cigars	<0.1–0.42 mg/m³	2, 1, 5
	38–170 m³ rooms	none	10 cigarettes, 9 cigars	0.13–1.04 mg/m³	2; 5
Total particulate matter	15–425 m³ homes	1–3	7–35 cigarettes	1.1–3.0 mg/m³	6, 7
	25 m³ chamber	none	4–24 cigarettes	2.28–16.65 mg/m³	3
Dimethylnitrosamine	4-m³ box, 20 m³ room	none	10–100 cigarettes	0.23–2.9 μg/m³	8
Acrolein	30–170 m³ rooms	none–2.4	5–150 cigarettes	0.02–0.20 ppm	5, 9
Acetaldehyde	38–170 m³ rooms	none–2.4	5–150 cigarettes	0.06–0.56 ppm	5
Formaldehyde	30 m³ box	none	5–10 cigarettes	0.23–0.46 ppm	9
NO	30 m³ box	none	5–10 cigarettes	0.19–0.36 ppm	9
NO$_2$	30 m³ box	none	5–10 cigarettes	0.02–0.04 ppm	9
Natural Conditions					
CO	Office, restaurant club, tavern, arena			2.5–28 ppm	10, 11, 12, 13, 14
	submarine, boat, autos, bus, airplane	none–20	4–150 cigarettes	3–33 ppm	15, 16, 17, 18
Nicotine	submarine, terminal, restaurant	—	Up to 150 cigarettes	1–35 μg/m³	15, 19
Total particulate matter	tavern, arena	none–6	—	0.15–0.98 mg/m³	11, 12
Particles	house	—	1 cigar	48 × 10⁶ particles/ft³	20
Benzopyrene	arena	—	—	0.0071–0.021 μg/m³	12, 21
Dimethylnitrosamine	bar	—	—	0.11–0.24 μg/m³	8
Respirable particulate matter (RP)	restaurants, sports arena, bowling alley	—	—	100–700 μg/m³	22

[a]From Chapter 11, *Smoking and Health—A Report of the Surgeon General*, HEW (1979).

[b]*Sources:* (1) Anderson and Dalhamn (1973); (2) Harke (1970); (3) Hoegg (1972); (4) Harke et al. (1974); (5) Harke et al. (1972); (6) Lawther and Commins, (1970); (7) McNall (1975); (8) Brunnemann and Hoffman (1978); (9) Weber et al. (1976); (10) Chappell and Parker (1977); (11) Cuddeback et al. (1976); (12) Elliott and Rowe (1975); (13) Harke (1974); (14) Sebben et al. (1977); (15) Cano et al. (1970); (16) Godin et al. (1972); (17) Harke and Peters (1974); (18) Seiff (1973); (19) Hinds and First (1975); (20) Lefcoe and Inculet, 1975; (21) Galuskinova (1964); (22) Repace and Lowrey (1980).

Table 2.7 Health Studies of Involuntary Smoking

Health Study	Number of Subjects (age)	Results	References
Occupationally exposed smokers and nonsmokers, 83% in professional, technical, or managerial positions (southern California)—spirometry	2100 (middle aged, 40's)	Nonsmokers exposed at work had lower values on lung function tests which are related to significantly reduced small airway function, FEF_{25-75} and FEF_{75-85}.	White and Froeb (1980)
British children of smoking parents—questionnaire	2205 (<5)	Parental smoking the cause of increased respiratory disease in first year of life: ~3 additional cases of pneumonia or bronchitis per 100 children for each adult smoker.	Colley et al. (1974); Leeder et al. (1976)
Children of smoking and nonsmoking parents—spirometry; questionnaire	444 (5–9)	FEF_{25-75} lower for children of smoking parents vs. children of nonsmoking parents.	Tager et al. (1979)
Children of smoking and nonsmoking Finnish mothers	12,000 (<5)	Children of smoking mothers (matched by marital status, maternal age, and socioeconomic status) had significantly higher morbidity (mostly respiratory disease) and were more likely to be hospitalized and for longer periods of time than children of nonsmoking mothers. Most pronounced effect in morbidity in first year of life.	Rantakallio (1978a, b)
Patients with angina; 15 cigarettes in 31-m^3 room in 2 hr—cardiovascular measurements	10 (40–60)	Increased measures of cardiac function and decreased duration to anginal pain.	Aronow (1978)
Gas–electric stove studies—spirometry; questionnaire	808 (6–7) 8120 (6–10)	More respiratory disease in homes with smoking parents.	Florey et al. (1979); Speizer et al. (1980)

Population	Number (age)	Findings	Reference
Nonsmoking Japanese wives	91,540 (≥40)	Wives of heavy smokers had significantly greater risk of developing lung cancer; age-occupation standardized annual mortality rates for lung cancer 8.7/100,000 for wives of occasional or nonsmokers; 14/100,000 for wives of exsmokers or those smoking ≤19 cigarettes/day; 18.1/100,000 for wives of those smoking ≥20 cigarettes/day. The relative risk of passive smoking was about $\frac{1}{3}$ to $\frac{1}{2}$ that of direct smoking.	Hirayama (1981)
Nonsmoking Greek women	189 (mean age = 62–63 years)	Statistically significant difference between cancer cases (40) and other patients (149) with respect to husbands' smoking habits. Relative risks of lung cancer were 2.4 for those with husbands who smoked <20 cigarettes/day and 3.4 for >20 cigarettes/day. Tentatively, the relative risk of passive smoking was 80% of that for direct smoking, but with broad confidence limits.	Trichopoulos et al. (1981)
Nonsmoking American women	176,739 (35–89)	Unadjusted lung cancer mortality ratios were 1.27 for women with husbands who smoked <20 cigarettes/day and 1.10 for those married to ≥20 cigarettes/day smokers. When compared to those with nonsmoking husbands, neither mortality ratio was statistically significant. When data adjusted for age, race, educational status, residence, and husband's occupational exposure to dust, fumes, or vapors the mortality ratios were 1.37 for those with husbands smoking <20 cigarettes/day and 1.04 for spouses of ≥20 cigarettes/day smokers, neither ratio being statistically significant.	Garfinkel (1981)

to reduce the time at which angina occurs after exercise (Aronow, 1978). Chronic exposure to tobacco smoke in the work environment has been found to significantly reduce small-airways function in nonsmokers (White and Froeb, 1980). Questions on involuntary smoking by workers are still more serious in view of the known synergism between voluntary smoking and occupational carcinogens such as asbestos and uranium (Selikoff et al., 1972; Archer et al., 1973). Table 2.7 summarizes some of the major findings from recent studies of involuntary smoking.

ASBESTOS AND FIBROUS PARTICLES

Asbestos is a generic term that applies to several naturally occurring, hydrated mineral silicates. The most common form in the United States is chrysotile ($3MgO \cdot 2SiO_2 \cdot 2H_2O$). Other types include amosite, crocidolite, tremolite, anthophyllite, and actinolite. Asbestos particles in the ambient air appear as fibers. About 68% of the asbestos produced in the United States is used in bonded form in the construction industry in products such as floor tiles, asbestos cements, and roofing felts and shingles (NIOSH, 1972). Another 6% is friable or in powder forms in insulation and acoustical products and asbestos cement powders. Up to 80,000 fibers/m^3 (>5 μm in length) have been found in British and American office and public buildings and residences (NIOSH, 1976). Occupational exposures have been reported up to the 10^8-10^9 fibers/m^3 range (NIOSH, 1972); NIOSH, 1976) and higher exposures may have occurred in the past.

Significantly increased risk of death from nonmalignant respiratory disease (often asbestosis) has been reported for workers in the insulation, textile, and other asbestos industries (Selikoff, 1975; Leman et al., 1980; Newhouse, 1969). In addition, all commercial forms of asbestos have been shown to be carcinogenic in man (Leman et al., 1980). Increased rates of bronchial cancer and pleural and peritoneal mesotheliomas have been associated with asbestos exposure of asbestos manufacturing, insulating, and shipyard workers and among asbestos miners (e.g., see Leman et al., 1980 for summary of studies). Many of these same populations also exhibited excess risk of the gastrointestinal tract. Household contact with work-derived asbestos dust has also been associated with increased mesothelioma risk (Anderson et al., 1976, 1979). In general there has been a 15-20 year latent period between asbestos exposure and incidence of cancer. In addition, the level of exposure which caused the disease is ordinarily not known.

There are presently no epidemiological studies linking glass or rock fibers, which are asbestos substitutes, with lung cancer. Workers occupationally exposed to airborne fibrous glass particles (1.8 μm median diameter by number) have exhibited a significant excess of nonmalignant respiratory disease when compared with the total white male Caucasian population of the United States (Bayliss et al., 1976). Past glass fiber production has tended to generate particles larger than 1 μm in diameter. Certain more recent operations discharge a much larger number in the submicron range, and there is some slight evidence for increases in malignant and nonmalignant respiratory tract disease in pilot plant workers exposed to these particles (Bayliss et al., 1976). However, there is presently not enough experience in working with submicron glass fibers to judge their hazard. There is a suggestion from animal studies that such fibers may contribute to disease development (Hill, 1977; Bayliss et al., 1976).

FORMALDEHYDE

Formaldehyde (HCHO) is an important industrial chemical used to produce synthetic urea- and phenol-formaldehyde resins. These resins are applied primarily as adhesives in making particleboard, fiberboard, plywood, and laminates. Urea-formaldehyde concentrates are also used in coating processes, paper products, and in making foams for thermal insulation. The textile industry uses HCHO in the production of creaseproof, crushproof, flame-resistant, and shrinkproof fabrics.

Formaldehyde is one of the reaction products of atmospheric photochemical smog. It is also present in tobacco smoke, emissions from combustion processes, and emanations from furniture, building materials, and textiles containing HCHO resins.

Burning of the eyes, lacrimation, and general irritation of the upper respiratory passages are the first signs experienced at HCHO concentrations in the 0.1–5 ppm range. The odor of formaldehyde is generally sensed at 1 ppm but some individuals can detect it at 0.05 ppm (NAS, 1981a). Concentrations of 10–20 ppm may produce coughing, tightening in the chest, a sense of pressure in the head, and palpitation of the heart. These symptoms may occur in susceptible persons at <5 ppm, and those with bronchial asthma may experience acute asthmatic attacks on exposure to 0.25–5 ppm (NAS, 1981a). Exposures above 50–100 ppm can cause serious injury such as collection of fluid in the lungs (pulmonary edema), inflammation of the lungs (pneumonitis), or death (NIOSH, 1980). The irritant effects on persons exposed to formaldehyde are summarized in Tables 2.8 and 2.9. Table 2.10 gives estimated irritation responses for populations exposed to HCHO. [It is well to note with respect to Table 2.10 that some persons are sensitized to HCHO which can lead to allergic reactions to subsequent exposures otherwise considered low (e.g., see Bardana, 1980).]

In December 1980, the National Institute for Occupational Safety and Health (NIOSH) and the Occupational Safety and Health Administration (OSHA) rec-

Table 2.8 Summary of Environmental and Occupational Formaldehyde Exposure Studies[a]

Concentration (ppb)	Exposure	Effects	References
20,000	Chamber (<1 min)	Discomfort, lacrimation	Barnes and Speicher (1942)
13,800	Chamber (30 min)	Eye and nose irritation	Sim and Pattle (1957)
500–10,000	Indoor residential air	Eye irritation, headaches, GI tract symptoms, skin problems, respiratory complaints	Sardinas et al. (1979)
4000–5000	Occupational (10–30 min)	Irritation, discomfort, lacrimation	Fassett (1963)
670–4820	Indoor residential air (infants)	Vomiting, diarrhea, lacrimation, nosebleed, rash	Wisconsin Division of Health (1978)
20–4150	Indoor residential air	Eye and upper respiratory tract irritation, headache, tiredness, nausea, diarrhea	Wisconsin Division of Health (1978)
900–2700	Occupational	Upper respiratory tract irritation, lacrimation, % of people experiencing eye irritation increased from 60 to 92% with increasing HCHO concentration	Blejer and Miller (1966)
300–2700	Occupational	Annoying odor, lacrimation, irritation of respiratory tract, disturbed sleep	Shipkovitz (1968)
30–1770	Indoor residential air	Drowsiness, nausea, headache, nose and respiratory tract irritation	Breysse (1981)
900–1600	Occupational	Intense eye irritation and itching; dry, sore throat; increased thirst; disturbed sleep	Morrill (1961)
250–1390	Occupational	Upper respiratory tract irritation, coughing, headaches	Kerfoot and Mooney (1975)
400–800	Occupational	Lowered $FEV_{1.0}/FVC$, upper respiratory tract irritation	Schoenberg and Mitchell (1975)
130–450	Occupational	Burning and stinging of eyes, nose, and throat; headache	Bourne and Seferian (1959)
196–448	Day care centers	Significantly higher frequency of abnormal tiredness, mucous membrane irritation, headache, menstrual irregularities, and use of analgetics; control group exposure varied from 41–90 ppb	Olsen and Dossing (1982)

[a]NAS (1980).

28

Table 2.9 Summary of Clinical Studies with Formaldehyde[a]

Concentration (ppb)	Duration of Exposure	Number of Subjects	% of Subjects Responding	Effects	References
30–3200[b]	35 min	33	45 36 19	No significant change in eye blinking rate Doubling of eye blinking rate Increases in eye blinking rate	Weber-Tschopp et al. (1977)
30–2100[b]	20 min	33	33 20 10 7	Doubling of eye blinking rate "Desire to leave the room" Medium eye irritation Strong odor, strong eye irritation	Weber-Tschopp et al. (1977)
1600	5 hr/day × 4 days	16	94	"Slight discomfort," conjunctival irritation, dryness of nose and throat	Andersen (1979)
830	5 hr/day × 4 days	16	94	"Slight discomfort," conjunctival irritation, dryness of nose and throat	Andersen (1979)
30–500[b]	5 min	33	11 3 2	Doubling of eye blinking rate "Desire to leave the room" Medium eye irritation	Weber-Tschopp et al. (1977)
420	5 hr/day × 4 days	16	31	"Slight discomfort," conjunctival irritation, dryness of nose and throat	Andersen (1979)
250	5 hr/day × 4 days	16	19	"Slight discomfort," conjunctival irritation, dryness of nose and throat	Andersen (1979)

[a]NAS (1980).
[b]Total exposure for 35 min at concentrations increasing from 30 to 3200 ppb.

Table 2.10   Predicted Irritation Responses of Humans Exposed to Airborne Formaldehyde[a]

Concentration (ppb)	% of Population Giving Indicated Response	Degree of Irritation[b]
1500–3000	20	7–10
	>30	5–7
500–1500	10–20	5–7
	>30	3–5
250–500	20	3–5
<250	<20	1–3

[a]NAS (1980).

[b]*Irritation Index* (scale derived from clinical effects noted in the literature):

10–Strong eye, nose, and throat irritation; great discomfort; strong odor.

7–Moderate eye, nose, and throat irritation; discomfort.

5–Mild eye, nose, and throat irritation; mild discomfort.

3–Slight eye, nose, and throat irritation; slight discomfort.

1–Minimal eye, nose, and throat irritation; minimal discomfort.

0–No effects.

ommended that formaldehyde be handled in the workplace as a potential occupational carcinogen. This judgment was based on the high incidence of a rare form of nasal cancer in test animals undergoing extended exposures to 2, 6, and 15 ppm HCHO. At the present time epidemiologic studies do not permit a definitive determination of the carcinogenic risk to humans (NIOSH, 1980).

SUSPENDED PARTICULATE MATTER

Suspended particulate matter is generally considered to consist of all airborne solid and low-vapor-pressure liquid particles less than a few hundred micrometers in diameter. This type of definition does not discriminate in terms of particle size (<5-μm diameter particles have a much higher probability of deposition in the tracheo-bronchial and alveolar spaces of the lung) or chemical composition. Health effect studies on populations undergoing nonoccupational exposures often have included consideration of sulfur dioxide as well as particles since these two pollutants are discharged together by many emission sources. Tables 2.11 and 2.12 summarize much of the environmental health information on these two pollutants (Ferris, 1978).

In general, the concentration of total suspended particulate matter (TSP), as measured by the high-volume particle sampler, is significantly lower indoors than outdoors. This observation also holds true for SO_2 and its particulate product,

Table 2.11   Summary of Effects of Sulfur Dioxide and Particulate Matter on Human Health: Short-Term Effects[a]

SO$_2$		Suspended Particulates	Effects
$\mu g/m^3$	ppm	($\mu g/m^3$)	
250	(0.095)	350[b]	Increased respiratory symptoms in patients with chronic bronchitis
722	(0.276)	350[b]	No change in pulmonary function of patient with chronic lung disease
200–300	(0.076–0.114)	230[b]	Decreased FEV$_{1.0}$
200	(0.076)	150	Increased frequency of asthma attacks

[a]Ferris (1978).
[b]Corrected from original data to TSP equivalents.

Table 2.12   Summary of Effects of Sulfur Dioxide and Particulates on Human Health: Long-Term Effects[a]

SO$_2$		Suspended Particulates	Effects
$\mu g/m^3$	ppm	($\mu g/m^3$)	
250	(0.095)	250[b]	Increased phlegm production
130	(0.05)	240[b]	Increased respiratory disease
120	(0.046)	180[b]	Increased respiratory illness and decreased pulmonary function
120	(0.046)	230[b]	Increased lower respiratory illness
90	(0.037)[c]	93	Decreased FVC, FEV$_{0.75}$[d]
23	(0.009)	110	Decreased FEV$_{0.75}$[d]
425–50	(0.162–0.019)	195–85	Increased lower respiratory disease morbidity
55	(0.021)[c]	180	Increased respiratory symptoms, decreased pulmonary function
37	(0.014)[c]	131	No effect
66	(0.025)[c]	80	No effect

[a]Ferris (1978).
[b]Corrected from original data to TSP equivalents.
[c]SO$_2$ equivalent calculated from lead peroxide data.
[d]FVC, forced vital capacity; FEV$_{0.75}$, forced expiratory volume at 0.75 seconds.

sulfate (SO_4^{2-}), neither of which have significant indoor sources (Dockery and Spengler, 1981a, b). However, there has been considerable interest in the respirable fraction of suspended particulate matter (RP), which is the particle mass <2.5 μm in diameter (e.g., Miller et al., 1979). Since RP is contained in cigarette smoke, consumer spray-products, and other indoor sources, the potential health implications are of concern. However, at present, such health information is not available.

RADON AND RADON DAUGHTERS

Radon-222 (Rn-222) is a noble gas decay product of radium-226 (Ra-226) which in turn is part of the decay chain of uranium-238 (U-238). Ra-222 has a half life ($t_{1/2}$) of 3.8 days, is an alpha emitter as are several of its daughters (polonium-218, $t_{1/2}$ = 3 min; polonium-214, $t_{1/2}$ = 1.6 $\times$ 10^{-4} sec), and is essentially inert to chemical reaction. Any substance containing uranium or radium is a source of Ra-222. (Since the $t_{1/2}$ for Ra-226 is over 1600 years, the Rn-222 production rate is essentially a constant.) Various types of soil (e.g., phosphate ores and mining wastes) and masonry building materials have been identified as sources (Guimond et al., 1979; UN, 1977). The average indoor-to-outdoor radiation dose ratio has been estimated at 1.3 for buildings constructed of granites, brick, and concrete, in contrast to a ratio of about 0.75 for wooden structures (UN, 1977).

The presence of Rn-222 and its daughter products have been identified as a major factor in the causation of lung cancer in uranium miners in the United States, Canada, and Czechoslovakia, and nonuranium miners in Sweden, Newfoundland, and the United Kingdom (Guimond et al., 1979; UN, 1977). Based on the epidemiological studies of miners, the mean induction rate for lung cancer for all ages has been estimated to be in the region of (200–450 $\times$ 10^{-6} rad^{-1} of radon-222 type radiation and is likely to be higher for males and females over 35 years of age (UN, 1977). (The rad, the common unit of absorbed radiation dose, is equal to 100 ergs/g in any medium.) The exposures reported for the mining studies are ordinarily two or three orders of magnitude higher than those expected for indoor settings (NAS, 1981b).

OZONE

Ozone is a pulmonary irritant that affects the mucous membranes, other lung tissues, and respiratory function (EPA, 1979a). Most of ambient ozone is produced as a result of the action of sunlight on nitrogen oxides and hydrocarbons which, in turn, are contained in exhaust gases from automobiles and other combustion processes. Indoor sources that may contribute significant amounts of ozone include copying machines and electrostatic air cleaners (Allen et al., 1978; Selway et al., 1980; Sutton et al., 1976). Health effects associated with low-level exposure to ozone are summarized in Table 2.13.

CARBON DIOXIDE

Carbon dioxide (CO_2) is produced by human metabolism and exhaled through the lungs. The amount of CO_2 produced is a function of food composition and the activity level of an individual. The amount of CO_2 normally exhaled by an adult with an activity level representative of an office worker is about 200 ml/min (0.0073 cfm) (Woods, 1980).

Exposure of healthy individuals for prolonged periods to 1.5% CO_2 apparently causes mild metabolic stress while exposure to 7-10% will produce unconsciousness within a few minutes (ACGIH, 1979). Exposure of nuclear submarine crews to 0.7-1.0% CO_2 demonstrated a consistent increase in respiratory minute volume and cyclic changes in the acid-base balance in blood. These biochemical changes may be related to CO_2 uptake and release in bone. This effect may cause reduction in bone density due to release of calcium (Schaefer, 1979; Tansey et al., 1979). Ventilation standards are normally set to maintain CO_2 indoor concentrations $\leqslant 0.5\%$, a level which appears not to adversely affect persons with normal health (ASHRAE, 1980).

ORGANIC COMPOUNDS

A great variety of organic materials have been identified in indoor air. These include aliphatic and aromatic hydrocarbons, chlorinated hydrocarbons, and various ketones and aldehydes (Johansson, 1978; Jarke et al., 1981). While some of these have been suggested as possible carcinogens (e.g., benzene and tetrachloroethylene) the actual health implications of most organics found in indoor air are not presently well defined.

VIABLE PARTICULATE MATTER

Pollen, bacteria, fungal and plant spores, and viruses are all associated with airborne particles. A common measurement of viable particles is called TVP (total viable particles) or CFP (colony forming particles). Ordinarily this measure reflects bacterial activity and does not include pollen or viruses and often also excludes fungal spores. In general, it appears that TVP concentrations inside do not

Table 2.13 Compilation of Results Reported in Human Studies Examining Ozone or Oxidant Exposure[a]

Concentration (ppm)	Exposure Duration (hours) (for Clinical Studies); Averaging Time (for Epidemiological Studies)	Pollutant Measured (O_3 = ozone; O_x = oxidant)	Reported Effects	References
0.01–0.30	Hourly average	O_3	Lung function parameters in about 25% of Japanese school children tested were significantly correlated with O_3 concentrations (over the range of 0.01–0.30 ppm) in the 2 hr prior to testing.	Kagawa and Toyama (1975); Kagawa et al. (1976)
0.03–0.30	Hourly average	O_x	Although significant correlation was observed between decreased athletic performance and O_x concentrations in the range of 0.03–0.30 ppm, the criteria document states that inspection of the data reveals no obvious relationship between performance and O_x values below 0.10–0.15 ppm.	Wayne et al. (1967)
0.10	2	O_3	Decreased O_2 pressure in arterialized blood, increased airway resistance observed using nonstandard measurement techniques.	von Nieding et al. (1976)
0.10–0.15	Probably daily maximum hourly average	O_x	Increased rates of respiratory symptoms and headache were reported by Japanese students on days when O_x concentrations exceeded 0.15 ppm as compared to days when O_x concentrations were less than 0.10 ppm.	Makino and Mizoguchi (1975)
0.15	1	O_3	Subjective symptoms of discomfort were observed by most subjects, and discernible but not statistically significant changes in respiratory patterns occurred while performing vigorous exercise.	DeLucia and Adams (1977)
0.20	3	O_3	Reduction in visual acuity (night vision) observed.	Lagerwerff (1963)
0.20–0.25	2	O_3	Asthmatic patients exposed under intermittent, light exercise conditions showed no statistically significant changes in respiratory function. Symptom scores increased slightly during O_3 exposures. Small but statistically significant blood biochemical changes occurred.	Linn et al. (1978)

34

(ppm)	Duration	Pollutant	Effects	Reference
0.25	2	O_3	Small changes in lung function were observed in 3 subjects performing intermittent, light exercise.	Hazucha (1973)
0.25	2 and 4	O_3	No lung function changes of note were observed in "reactive" subjects (who had histories of cough, chest discomfort, or wheezing associated with air pollution or allergy) while performing intermittent, light exercise.	Hackney et al. (1975a, b, c)
0.25	Daily maximum hourly average	O_x	The average number of asthma patients having attacks was statistically significantly elevated on days when O_x levels exceeded 0.25 ppm.	Schoettlin and Landau (1961)
0.25	0.5-1	O_3	Blood samples of exposed subjects had increased rates of sphering of red blood cells.	Brinkman et al. (1964)
0.28	Daily maximum instantaneous (2-min) average	O_x	Although the reported results are inconclusive, EPA's examination of the evidence presented suggests exacerbation of asthma when O_x levels are above 0.28 ppm.	Kurata et al. (1976)
0.30	1	O_3	Subjective symptoms of discomfort and statistically significant changes in pulmonary function were observed in subjects undergoing vigorous exercise.	DeLucia and Adama (1977)
0.30	Daily maximum hourly average	O_x	Increased rates of cough, chest discomfort, and headache were observed in student nurses on days when the O_x concentrations exceeded 0.30 ppm.	Hammer et al. (1974)
0.37	2	O_3	Discomfort symptoms and significant changes in lung function were observed in subjects undergoing intermittent, light exercise.	Hazucha (1973); Folinsbee et al. (1975); Silverman et al. (1976)
0.37 / 0.37	2 / 2	O_3 / SO_2	Exposure to O_3 and SO_2 together produced changes in lung function substantially greater than the sum of the separate effects of the individual pollutants.	Hazucha and Bates (1975)
0.37 / 0.37	2 / 2	O_3 / SO_2	The observed O_3–SO_2 interactive effect on lung function was considerably smaller than that seen by Hazucha and Bates. The authors concluded that the earlier study probably more nearly simulated a smog episode in regions having high oxidant and sulfur pollution.	Bell et al. (1977)

[a]EPA (1979a).

35

follow outdoor concentrations but are more closely related to living conditions and indoor activity (Yocum et al., 1977). It has been estimated, based on measurements in schools, hospitals, and residences, that humans live in air with "bioburdens" of from 20 CFP/m^3 to over 700 CFP/m^3 without apparent ill effects (Berk et al., 1980). Spore levels are generally lower indoors than outdoors as are pollen grains (Yocum et al., 1977).

Air conditioners and cool-mist humidifiers have been identified as devices where pathogenic organisms may concentrate and later be released as concentrated viable aerosols (NAS, 1981b). *Legionellapneumophilia* (Legionnaires' disease) and *Acinetobacter* infection have been linked to such sources, but without quantitation of the exposure–effect relationship (e.g., Cordes et al., 1980; Smith, 1977). Such a relationship was developed for a measles epidemic in 1974 in Rochester, New York. The concentration of infectious particles which produced 28 secondary cases of the disease was estimated at 1 particle/5.17 m^3 of air (Riley et al., 1978).

REFERENCES

ACGIH (1979) *Documentation of the threshold limit values*, pp. 69–70, American Conference of Governmental Industrial Hygienists, Cincinnati, Ohio.

Allen, R. J., Wadden, R. A., and Ross, E. D. (1978). Characterization of potential indoor sources of ozone. *Am. Ind. Hyg. Assoc. J.* **39**:466–471.

Andersen, I. (1979). Formaldehyde in the indoor environment–health implications and the setting of standards. In: *Indoor Climate: Effects on Humans Comfort, Performance and Health in Residential, Commercial, and Light Industry Buildings. Proceedings of the First International Indoor Climate Symposium*, P. O. Fanger and O. Valbjoru, Eds., August 30 to September 1, 1978, Danish Building Research Institute, pp. 65–77. Discussion pp. 77–87.

Anderson, E. W., Andelman, R. J., Strauch, J. M., Fortuin, N. J., and Knelson, H. H. (1973). Effect of low-level carbon monoxide exposure on onset and duration of angina pectoris: A study on 10 patients with ischemic heart disease. *Ann. Internal Med.* **79**:46–50.

Anderson, G. and Dalhman, T. (1973). The risks to health of passive smoking. *Lakartidningen* **70**:2833–2836, August 15.

Anderson, H. A., Lilis, R., Daum, S. M., Fischbein, A. S., and Selikoff, I. J. (1976). Household-contact asbestos neoplastic risk. *Ann. N.Y. Acad. Sci.* **271**:311–323.

Anderson, H. A., Lilis, R., Daum, S., Fischbein, A., and Selikoff, I. J. (1979). Asbestosis among household contacts of asbestos factory workers. *Ann. N.Y. Acad. Sci.* **330**:387–399.

Archer, V. E., Wagoner, J. K., and Lurdin, F. E. (1973). Uranium mining and cigarette smoking effects on man. *J. Occup. Med.* **15**:204–211.

Aronow, W. S., Harris, C. N., Isbell, M. W., Rokaw, S. N., and Imparato, P. (1972). Effects of freeway travel on angina pectoris. *Ann. Internal Med.* **77**:669–676.

Aronow, W. S. and Isbell, M. W. (1973). Carbon monoxide effect on exercise-induced angina pectoris. *Ann. Internal Med.* **79**:392–395.

Aronow, W. S., Stemmer, E. A., and Isbell, M. W. (1974a). Effect of carbon monoxide exposure on intermittant claudication. *Circulation* **49**:415–417.

Aronow, W. S., Cassidy, J., Vangrow, J. S., March, H., Kern, J. C., Goldsmith, J. R., Khemka, M., Pagano, J., and Vawter, M. (1974b). Effects of cigarette smoking and breathing carbon monoxide on cardiovascular hemodynamics on anginal patients. *Circulation* **50**:340–347.

Aronow, W. S. and Cassidy, J. (1975). Effect of carbon monoxide on maximal treadmill exercise: A study in normal persons. *Ann. Internal Med.* **83**:496–499.

Aronow, W. S., Ferlinz, J., and Glauser, F. (1977). Effects of carbon monoxide on exercise performance in chronic obstructive pulmonary disease. *Am. J. Med.* **63**:904–908.

Aronow, W. S. (1978). Effect of passive smoking on angina pectoris. *N. Engl. J. Med.* **299**(1):21–24.

ASHRAE (1980). *Standards for ventilation required for minimum acceptable indoor air quality.* American Society of Heating, Refrigerating and Air-Conditioning Engineers, ASHRAE 62-73R, New York.

Bardana, E. J. (1980). Formaldehyde: Hypersensitivity and irritant reactions at work and in the home. *Immunol. Allergy Pract.* **11**:11–23.

Barnes, E. C. and Speicher, H. W. (1942). The determination of formaldehyde in air. *J. Ind. Hyg. Toxicol.* **24**:10–17.

Bayliss, D. L., Dement, J. M., Wagoner, J. K., and Blejer, H. P. (1976). Mortality patterns among fibrous glass production workers. *Ann. N.Y. Acad. Sci.* **271**:324–335.

Bell, K. A., Linn, W. S., Hazucha, M., Hackney, J. D., and Bates, D. V. (1977). Respiratory effects of exposure to ozone plus sulfur dioxide in southern Californians and eastern Canadians. *Am. Ind. Hyg. Assoc. J.* **38**:696–706.

Berk, J. V., Boyan, T. A., Brown, S. R., Ko, I., Koonce, J. F., Loo, B. W., Pepper, J. H., Robb, A. W., Strong, P. C., Turiel, I., and Young, R. A. (1980). Field monitoring of indoor air quality. In: *1979 Annual Report of the Energy and Environment Division*, Lawrence Berkeley Laboratory, University of California, Report No. LBL 11650.

Blejer, H. P. and Miller, B. H. (1966). *Occupational health report of formaldehyde concentrations and effects on workers at the Bayly Manufacturing Company, Visalia, Cal.*, Study Report S-1806. Bureau of Occupational Health, Department of Public Health, State of California Health and Welfare Agency, Los Angeles.

Bourne, H. G., Jr. and Seferian, S. (1959). Formaldehyde in wrinkle-proof apparel produces . . . tears for milady. *Ind. Med. Surg.* 28:232–233.

Breysse, P. A. (1981). The health cost of "tight" homes. *Am. J. Med. Assoc.* 245:267–268.

Brinkman, R. H., Lamberts, H. B., and Veninga, T. S. (1964). Radiomimetric toxicity of ozonated air. *Lancet* 1(7325):133–136.

Brunnemann, K. D. and Hoffmann, D. (1978). Chemical studies on tobacco smoke. LIX. Analysis of volatile nitrosamines in tobacco smoke and polluted indoor environments. In: *Environmental Aspects of N-Nitroso Compounds*, F. A. Walker, M. Castegnaro, L. Griciute, and R. E. Lyle, Eds., Lyon (IARC Scientific Publication No. 19), pp. 343–356.

Burgess, W., DiBerardinis, L., and Speizer, F. E. (1973). Exposure to automobile exhaust. III. An Environmental Assessment. *Arch. Environ. Health* 26:325–329.

Cano, J. P., Catalin, J., Badre, R., Dumas, C., Viala, A., and Guillerme, R. (1970). Determination de la nicotine par chromatographie en phase gazeuse. II.-Applications (Determination of nicotine by chromatography in the gaseous phase. II-Applications). *Ann. Pharmaceut. Fran.* 28(11):633–640.

Chappell, S. B. and Parker, R. J. (1977). Smoking and carbon monoxide levels in enclosed public places in New Brunswick. *Can. J. Public Health* 68:159–161.

Coburn, R. F., Blakemore, W. S., and Forster, R. E. (1963). Endogenous carbon monoxide production in man. *J. Clin. Invest.* 42:1172–1178.

Coburn, R. F., Forster, R. E., and Kane, P. B. (1965). Considerations of the physiological variables that determine the blood carboxyhemoglobin concentration in man. *J. Clin. Invest.* 44:1899–1910.

Cohen, C. A., Hudson, A. R., Clausen, J. L., and Knelson, J. H. (1972). Respiratory symptoms, spirometry, and oxidant air pollution in nonsmoking adults. *Am. Rev. Resp. Disease* 105:251–261.

Colley, J. R. T., Holland, W. W., and Corkhill, R. T. (1974). Influence of passive smoking and parental phlegm on pneumonia and bronchitis in early childhood. *Lancet* 2(7888):1031–1034, November 2.

Cordes, L. G., Fraser, D. W., Skaliy, P., Perlino, C. A., Elsea, W. R., Mallison, G. F., and Hayes, P. S. (1980). Legionnaires disease outbreak at an Atlanta, Georgia, Country Club: Evidence for spread from an evaporative condenser. *Am. J. Epidemiol.* 111:425–431.

Cuddeback, J. E., Donovan, J. R., and Burg, W. R. (1976). Occupational aspects of passive smoking. *Am. Ind. Hyg. Assoc. J.* 37(5):263–267.

DeLucia, A. J. and Adams, W. C. (1977). Effects of O_3 inhalation during exercise of pulmonary function and blood biochemistry. *J. Appl. Physiol. Respirat. Environ. Exercise Physiol.* 43(1):75–81.

Dockery, D. W. and Spengler, J. D. (1981a). Personal exposure to respirable particulates and sulfates. *J. Air Pollut. Control Assoc.* 31:153–159.

Dockery, D. W. and Spengler, J. D. (1981b). Indoor-outdoor relationships of respirable sulfates and particles. *Atmos. Environ.* 15:335–343.

Elliott, L. P. and Rowe, D. R. (1975). Air quality during public gatherings. *J. Air Pollut. Control Assoc.* 25:635–636.

EPA (1976). *Scientific and technical data base for criteria and hazardous pollutants, 1975 ERC/RTP Review.* U.S. Environmental Protection Agency, EPA-600/1-76-023, Research Triangle Park, North Carolina.

EPA (1979a). Revisions to the National Ambient Air Quality Standards for Photochemical Oxidant. *Fed. Reg.* 44:8201–8233.

EPA (1979b) *Air quality criteria for carbon monoxide.* U.S. Environmental Protection Agency, EPA-600/8-79-022, Washington, D.C.

EPA (1980). *Carbon* monoxide: Proposed revisions to the National ambient air quality standard and announcement of public meetings. *Fed. Reg.* 45:55066–55084.

EPA (1981). *Air Quality Criteria for Oxides of Nitrogen* (in press). U.S. Environmental Protection Agency, Environmental Criteria and Assessment Office, Washington, D.C.

Fassett, D. W. (1963). Aldehydes and acetals. In: *Industrial Hygiene and Toxicology* F. A. Patty, Ed., 2nd rev. ed., Vol. II, pp. 1959–1989, Interscience, New York.

Ferris, B. G. (1978). Health effect of exposure to low levels of regulated pollutants—A critical review. *J. Air Pollut. Control Assoc.* 28:482–497.

Florey, C. duV., Melia, R. J. W., Chinn, S., Goldstein, B. D., Brooks, A. G. F., John, H. H., Craighead, I. B., and Webster, X. (1979). The relation between respiratory illness in primary school children and the use of gas for cooking. III-Nitrogen dioxide, respiratory illness and lung infection. *Int. J. Epid.* 8:347–353.

Folinsbee, L. J., Silverman, F., and Shephard, R. L. (1975) Exercise responses following ozone exposure. *J. Appl. Physiol.* 38:996–1001.

Galuskinova, V. (1964). 3,4-Benzpyrene determination in the smoky atmosphere of social meeting rooms and restaurants. A contribution to the problem of the noxiousness of so-called passive smoking. *Neoplasma* 11(5):465–468.

Garfinkel, L. (1981). Time trends in lung cancer mortality among non-smokers and a note on passive smoking. *J. Natl. Cancer Inst.* 66:1061–1066.

Godin, G., Wright, G., and Shephard, R. J. (1972). Urban exposure to carbon monoxide. *Arch. Environ. Health* 25(5):305–313, November.

Goldstein, B. D., Melia, R. J. W., Chinn, S., Florey, C. duV., Clark, D., and John, H. H. (1979). The relation between respiratory illness in primary school children and the use of gas for cooking. II—Factors effecting nitrogen dioxide levels in the home. *Int. J. Epid.* 8:339–346.

Guimond, R. J., Ellett, W. H., Fitzgerald, J. E., Windham, S. T., and Cuny, P. A. (1979). *Indoor radiation exposure due to radium-226 in Florida phosphate lands,* U.S. EPA Report EPA-502/4-78-013.

Hackney, J. D., Linn, W. S., Buckley, R. D., Pedersen, E. E., Karuze, S. K., Law, D. C., and Fischer, D. A. (1975a). Experimental studies on human health effects of air pollutants. I. Design considerations. *Arch. Environ. Health* **30**:373-378.

Hackney, J. D., Linn, W. S., Mohler, J. G., Pedersen, E. E., Breisacher, P., and Russo, A. (1975b). Experimental studies on human health effects of air pollutants. II. Four-hour exposure to ozone alone and in combination with other pollutant gases. *Arch. Environ. Health* **30**:379-384.

Hackney, J. D., Linn, W. S., Law, D. C., Karuza, S. K., Greenberg, H., Buckley, R. D., and Pedersen, E. E. (1975c). Experimental studies on human health effects of air pollution. III. Two-hour exposure to ozone alone and in combination with other pollutant gases. *Arch. Environ. Health* **30**:385-390.

Hammer, D. J., Hasselblad, V., Portnoy, B., and Wehrle, P. F. (1974). The Los Angeles student nurse study. *Arch. Environ. Health* **28**:255-260.

Harke, H.-P. (1970). The problem of "passive smoking." *Muenchener Med. Wochenschr.* **112**(5):2328-2334, December 18.

Harke, H.-P. (1974). Zum Problem des Passivrauchens. I. Ueber den Einfluss des Rauchens auf die CO-Konzentration in Bueroraeumen (The problem of passive smoking. I. The influence of smoking on the CO concentration in office rooms). *Int. Arch. Arbeitsmed.* **33**(3):199-206.

Harke, H.-P., Baars, A., Frahm, B., Peters, H., and Schultz, C. (1972). The problem of passive smoking. The concentration of smoke constituents in the air of large and small rooms as a function of the number of cigarettes smoked and of time. *Int. Arch. Arbeitsmed.* **29**:323-339.

Harke, H.-P., Liedl, W., and Denker, D. (1974). Zum Problem des Passivrauchens. II. Untersuchungen Ueber den Kohlenmonoxidgehalt der Luft im Kraftfahrzeug durch das Rauchen von Zigaretten (The problem of passive smoking. II. Investigations of CO level in the automobile after cigarette smoking). *Int. Arch. Arbeitsmed.* **33**(3):207-220.

Harke, H.-P. and Peters, H. (1974). Zum Problem des Passivrauchens. III. Ueber den Einfluss des Rauchens auf die CO-Konzentration im Kraftfahrzeug bei Fahrten im Stadtgebiet (The problem of passive smoking. III. The influence of smoking on the CO concentration in driving automobiles). *Int. Arch. Arbeitsmed.* **33**(3):221-229.

Harlap, S. and Davies, A. M. (1974). Infant admissions to hospital and maternal smoking. *Lancet* **1**(7857):529-532, March 30.

Hazucha, M. (1973). Effects of ozone and sulfur dioxide on pulmonary function in man. Ph.D. thesis, McGill University, Montreal, Canada.

Hazucha, M. and Bates, D. V. (1975). Combined effect of ozone and sulfur dioxide on human pulmonary function. *Nature* **257**(5521):50-51.

HEW (1979). *Smoking and Health—A Report of the Surgeon General.* Department of Health, Education and Welfare, Pub. No. (PHS) 79-50066.

Hill, T. W. (1977). Health aspects of man-made mineral fibers. *Ann. Occup. Hyg.* **20**:161-173.

Hinds, W. C. and First, M. W. (1975). Concentrations of nicotine and tobacco smoke in public places. *N. Engl. J. Med.* **292**(16):844–845. April 17.

Hirayama, T. (1981). Non-smoking wives of heavy smokers have a higher risk of lung cancer: A study from Japan. *Brit. Med. J.* **282**:183–185.

Hoegg, U. R. (1972). Cigarette smoke in closed spaces. *Environ. Health Perspec.* **2**:117–128, October.

Jarke, F. H., Dravnieks, A., and Gordon, S. M. (1981). Organic contaminants in indoor air and their relation to outdoor contaminants. *ASHRAE Trans.* **87**(Part 1):153–166.

Johansson, I. (1978). Determination of organic compounds in indoor air with potential reference to air quality. *Atmos. Environ.* **12**:1371–1377.

Kagawa, J. and Toyama, T. (1975). Photochemical air pollution: Its effect on respiratory function of elementary school children. *Arch. Environ. Health* **30**:117–122.

Kagawa, J., Toyama, T., and Nakaza, M. (1976). Pulmonary function tests in children exposed to air pollution. In: *Clinical Implications of Air Pollution Research*, A. J. Finkel and W. C. Duel, Eds., Publishing Sciences Group, Inc., Acton, Mass., pp. 305–320.

Keller, M. D., Lanese, R. R., Mitchell, R. I., and Cote, R. W. (1979a). Respiratory illness in households using gas and electric cooking. I. Survey of incidence. *Environ. Res.* **19**:495–503.

Keller, M. D., Lanese, R. R., Mitchell, R. I., and Cote, R. W. (1979b). Respiratory illness in households using gas and electric cooking. II. Symptoms and objective findings. *Environ. Res.* **19**:504–515.

Kerfoot, E. J. and Mooney, T. F., Jr. (1975). Formaldehyde and paraformaldehyde study in funeral homes. *Am. Ind. Hyg. Assoc. J.* **36**:533–537.

Kerr, H. D., Kulla, T. J., McIlhany, M. L., and Swidersky, P. (1979). Effect of nitrogen dioxide on pulmonary function in human subjects: An environmental chamber study. *Environ. Res.* **19**:392–404.

Kurata, J. H., Glousky, M. M. Newcomb, R. L., and Easton, J. G. (1976). A multifactorial study of patients with asthma. Part 2. Air pollution, animal dander and asthma symptoms. *Ann. Allergy* **37**:379–409.

Lagerwerff, J. M. (1963). Prolonged ozone inhalation and its effects on visual parameters. *Aerospace Med.* **34**:479–489.

Lawther, P. J. and Commins, B. T. (1970). Cigarette smoking and exposure to carbon monoxide. *Ann. N.Y. Acad. Sci.* **174**:135–147, October 5.

Leeder, S. R., Corkhill, R., Irwig, L. M., Holland, W. W., and Colley, J. R. T. (1976). Influence on family factors on the incidence of lower respiratory illness during the first year of life. *Brit. J. Prevent. Social Med.* **30**(4):203–212, December.

Lefcoe, N. M. and Inculet, I. I. (1975). Particulates in domestic premises. II. Ambient levels and indoor–outdoor relationships. *Arch. Environ. Health* **30**(12):565–570, December.

Lemen, R. A., Dement, J. M., and Wagoner, J. K. (1980). Epidemiology of asbestos-related diseases. *Environ. Health Prespect.* 34:1–11.

Linn, W. S., Hackney, J. D., Pedersen, E. E., Breisacher, P., Patterson, J. V., Mulry, C. A., and Coyle, J. F. (1976). Respiratory function and symptoms in urban office workers in relation to oxidant air pollution exposure. *Am. Rev. Resp. Disease* 114:477–483.

Linn, W. S., Buckley, R. D., Spier, C. E. Blessey, R. L., Jones, M. P., Fischer, D. A., and Hackney, J. D. (1978). Health effects of ozone exposure in asthmatics. *Am. Rev. Resp. Dis.* 117:835–843.

Makino, K. and Mizoguchi, I. (1975). Symptoms caused by photochemical smog. *Jpn. J. Public Health* 22(8):421–430.

McNall, P. E. (1975). Practical methods of reducing airborne contaminants in interior spaces. *Arch. Environ. Health* 30(11):552–556, November.

Melia, R. J. W., Florey, C. duV., Altman, D. S., and Swan, A. V. (1977). Association between gas cooking and respiratory disease in children. *Brit. Med. J.* 2:149–152.

Melia, R. J. W., Florey, C. duV., and Chinn, S. (1979). The relation between respiratory illness in primary school children and the use of gas for cooking. I. Results from a national survey. *Int. J. Epid.* 8:333–338.

Miller, F. J., Gardner, D. E., Graham, J. A., Lee, R. E., Jr., Wilson, W. E. and Bachman, J. D. (1979). Size considerations for establishing a standard for inhalable particle. *J. Air Pollut. Control Assoc.* 29:610–615.

Mitchell, R. I., Williams, R., Cote, R. W., Lanese, R. R. and Keller, M. D. (1974). Household survey of the incidence of respiratory disease in relation to environmental pollutants. *WHO Symposium Proceedings: Recent Advance in the Assessment of the Health Effects of Environmental Pollutants*, Paris, June 24–28.

Morrill, E. E., Jr. (1961). Formaldehyde exposure from paper process solved by air sampling and current studies. *Air Cond. Heat. Vent.* 58(7): 94–95.

NAS (1977a). *Carbon Monoxide.* National Academy of Sciences, Washington, D.C.

NAS (1977b). *Nitrogen Oxides.* National Academy of Sciences, Washington, D.C.

NAS (1980). *Formaldehyde–An assessment of its health effects.* National Academy of Sciences Report by the Committee on Toxicology to the Consumer Product Safety Commission, Washington, D.C., March.

NAS (1981a). *Formaldehyde and Other Aldehydes.* National Academy of Sciences, Washington, D.C.

NAS (1981b). *Indoor Pollutants.* National Academy of Sciences, Washington, D.C.

Newhouse, M. L. (1969). A study of the mortality of workers in an asbestos factory. *Brit. J. Ind. Med.* 26:294–301.

NIOSH (1972). *Occupational exposure to asbestos.* National Institute for Occupational Safety and Health, HSM 72-10267.

NIOSH (1976). *Revised recommended asbestos standard.* National Institute for Occupational Safety and Health, DHEW (NIOSH) 77-169.

NIOSH (1980). *Formaldehyde: Evidence of carcinogenicity.* Joint National Institute of Occupational Safety and Health/Occupational Safety and Health Administration Bulletin, No. 34, December 23.

Olsen, J. H. and Dossing, M. (1982). Formaldehyde induced symptoms in day care centers. *Am. Ind. Hyg. Assn. J.* **43**:366–370.

Orehek, J., Massari, J. P., Gayrard, P., Grimaud, C., and Charpin, J. (1976). Effect of short-term, low-level nitrogen dioxide exposure on bronchial sensitivity of asthmatic patients. *J. Clin. Invest.* **57**:301–307.

Peterson, J. E. and Stewart, R. D. (1975). Predicting the carboxyhemoglobin levels resulting from carbon monoxide exposures. *J. Appl. Physiol.* **39**:633–638.

Peterson, J. E. (1977). *Industrial Health.* Prentice-Hall, Englewood Cliffs, N.J.

Rantakallio, P. (1978a). Relationship of maternal smoking to morbidity and mortality in the child up to the age of five. *Acta Paeditr. Scand.* **67**:621–631.

Rantakallio, P. (1978b). The effect of maternal smoking on birthweight and the subsequent health of the child. *Early Human Dev.* **2**:371–382.

Repace, J. L. and Lowrey, A. H. (1980). Indoor air pollution, tobacco smoke, and public health. *Science* **208**:464–472, May 2.

Riley, E. C., Murphy, G., and Riley, R. L. (1978). Airborne spread of measles in a suburban elementary school. *Am. J. Epi.* **107**:421–432.

Sardinas, A. V., Most, R. S., Giulietti, M. A., and Honcher, P. (1979). Health effects associated with urea-formaldehyde foam insulation in Connecticut. *J. Environ. Health* **41**:270–272.

Schaeffer, K. E. (1979). Editorial summary (preventive aspects of submarine medicine). *Undersea Biomedical. Res.* **6**(Suppl.): S-7 to S-14.

Schoenberg, J. B. and Mitchell, C. A. (1975). Airway disease caused by phenolic (phenol-formaldehyde) resin exposure. *Arch. Environ. Health* **30**:574–577.

Schoettlin, C. E. and Landau, E. (1961). Air pollution and asthmatic attacks in the Los Angeles area. *Public Health Repts.* **76**:545–548.

Sebben, J., Pimm, P., and Shephard, R. J. (1977). Cigarette smoke in enclosed public facilities. *Arch. Environ. Health* **32**(2):53–58, March/April.

Seiff, H. E. (1973). *Carbon monoxide as an indicator of cigarette-caused pollution levels in intercity buses.* U.S. Department of Transportation, Federal Highway Administration, Bureau of Motor Carrier Safety, April, 11 pp.

Selikoff, I. J. Nicholson, W. J., and Langer, A. M. (1972). Asbestos air pollution. *Arch. Environ. Health* **25**:1–3.

Selikoff, I. J. (1975). Asbestos disease in the United States 1918-1975. Presented at the Conference on Asbestos Disease, Rouen, France, October 27.

Selway, M. D., Allen, R. J., and Wadden, R. A. (1980). Ozone emissions from photocopying machines. *Am. Ind. Hyg. Assn. J.* **41**:455–459.

Shipkovitz, H. D. (1968). *Formaldehyde vapor emissions in the permanent-press fabrics industry.* Report No. TR-52. Environmental Control Administration, Environmental Health Service, Consumer Protection and Environmental Health Service, Public Health Service, U.S. DEW, Cincinnati.

Silverman, F., Folinsbee, L. J., Barnard, J., and Shephard, R. J. (1976). Pulmonary function changes in ozone—Interaction of concentration and ventilation. *J. Appl. Physiol.* 41(6):859–864.

Sin, V. M. and Pattle, R. E. (1959). Effect of possible smog irritants on human subjects. *J. Am. Med. Assoc.* 165:1908–1913.

Smith, P. W. (1977). Room humidifiers as the source of *Acinetobacter* infections. *J. Am. Med. Assoc.* 237:795–797.

Speizer, F. E. and Ferris, B. G., Jr. (1973). Exposure to automobile exhaust. II. Pulmonary function measurements. *Arch. Environ. Health* 26(6):319–324.

Speizer, F. E., Ferris, B. G., Bishop, Y. M. M., and Spengler, J., (1980). Respiratory disease rates and pulmonary function in children associated with NO_2 exposure. *Am. Rev. Resp. Dis.* 121:3–10.

Sutton, D. J., Nodolf, K. M., and Makino, K. K. (1976). Predicting ozone concentrations in residential structures. *ASHRAE J.*, pp. 21–26, September.

Tager, I. B., Weiss, S. T., Rosner, B., and Speizer, F. E. (1979). Effect of parental cigarette smoking on the pulmonary function of children. *Am. J. Epid.* 110:15–26.

Tansey, W. A., Wilson, J. M., and Schaefer, K. E. (1979). Analysis of health data from 10 years of Polaris submarine patrols. *Undersea Biomedical Res.* 6(Suppl.): S-217 to S-246.

Trichopoulos, D., Kalandidi, A., Sparros, L., and MacMahon, B. (1981). Lung cancer and passive smoking. *Int. J. Cancer* 27:1–4.

UN (1977). *Sources and effects of ionizing radiation.* United Nations Scientific Committee on the Effects of Atomic Radiation 1977 Report to the General Assembly, with Annexes, United Nations, New York.

von Nieding, G., Wagner, H. M., Krekeler, H., Smidt, U., and Muysers, K. (1971). Minimum concentrations of NO_2 causing acute effects on the respiratory gas exchange and airway resistance in patients with chronic bronchitis. *Int. Arch. Arbeitsmed.* 27:338–348.

von Nieding, G., Krekeler, H., Fuchs, R., Wagner, H. M., and Koppenhagen, K. (1973). Studies of the acute effect of NO_2 on lung function: Influence on diffusion, perfusion and ventilation in the lungs. *Int. Arch. Arbeitsmed.* 31:61–72.

von Nieding, A. H., Wagner, H. M., Löellgen, H. L., and Krekeler, H. (1976). Presented at the VDI Kommission Reinhaltung der Luft Colloqium on Ozone and Related Substances in Photochemical Smog, Duesseldorf, West Germany, September 22–24.

Wayne, W. S., Wehrle, P. F., and Carroll, R. E. (1967). Pollution and athletic performance. *J. Am. Med. Assoc.* 199(12):901–904.

Weber, A., Fischer, T., Sancin, E., and Grandjean, E. (1976). La pollution de l'air par la fumee de cigarettes: Effets physiologiques et irritations (Air pollution due to cigarette smoke: Physiological and irritating effects). *Sozial- und praeventiv-medizin/Medecine Sociale et Preventive* **21**(4):130–132, July/August.

Weber-Tschopp, A., Fischer, T., and Grandjean, E. (1977). Irritating effects of formaldehyde on men. (In German, asbstract in English). *Int. Arch. Occup. Environ. Health* **39**:207–218.

White, J. R. and Froeb, H. F. (1980). Small airways dysfunction in non-smokers chronically exposed to tobacco smoke. *N. Engl. J. Med.* **302**:720–723.

Wisconsin Division of Health (1978). Bureau of Prevention, Section of Environmental Epidemiology. Statistics of particle board related formaldehyde cases through December 15, 1978, 4 pp.

Woods, J. E. (1980). Environmental implications of conservation and solar space heating. Energy Research Institute, Iowa State Univ., Ames, Iowa, BEUL 80-3. Meeting of the New York Academy of Sciences, New York, January 16.

Yocum, J. E., Coté, W. A., and Benson, F. B. (1977). Effects of indoor air quality. In: *Air Pollution*, A. E. Stern, Ed., Vol. II, 3rd ed., Academic Press, New York.

3

OUTDOOR CONTRIBUTIONS

A major component in determining indoor air pollution levels is often the outdoor concentration. Some of the materials initially in ambient air may be removed or destroyed on entering enclosed areas. For example, environmental control systems such as air conditioning and makeup air ventilation filters may reduce the eventual contribution of pollutants like TSP, and ambient ozone decays rapidly on a variety of air-entry surfaces. But outdoor levels always need to be considered when ventilation requirements are being determined to prevent unhealthful exposures.

Extensive ambient air data are available for many American areas for CO, TSP, NO_2, and ozone. Average annual concentrations are probably appropriate values to use given the philosophical approach that aside from specialized requirements (e.g., medical intensive care units, biological clean rooms, rare book collections), indoor areas are not ordinarily designed as pollution-control devices. Recent outdoor concentrations of TSP, CO, NO_2, and ozone for a number of American cities are given in Table 3.1.

Outdoor data for other materials are much more limited. Typical asbestos concentrations are given in Table 3.2. Formaldehyde concentrations for four New Jersey sites based on continuous measurements are given in Table 3.3 (Cleveland et al., 1977). Some of these data are also displayed in Figure 3.1. Data from a California site are shown in Figure 3.2 (Tuazon et al., 1980). Normally encountered outdoor Ra-222 levels in the United States are in the 0.04–1 pCi/liter range (NCRP, 1975), although concentrations over phosphate land tracts tend to be slightly higher (0.5–1.5 pCi/liter) (Windham et al., 1978). Carbon dioxide concentrations in the ambient environment are typically 300–600 ppm, with the higher value being appropriate for extensive urban areas with a multiplicity of combustion sources. Ambient concentrations of a variety of organic materials are given in Table 3.4 (Singh et al., 1981). Total viable particle concentrations in urban environments are typically about 150 organisms/m^3 (e.g., see Scheff et al., 1981).

Table 3.1 Annual Median Concentrations for TSP, NO_2, O_3, and CO for 1979[a]

	Concentration			
	$\mu g/m^3$			mg/m^3
Location	TSP (annual average)[b]	NO_2 (1-hr average)	O_3 (1-hr average)	CO (1-hr average)
Baltimore	43–102	45	20	1.5
Boston	67	75	–	3.5
Burbank, Calif.	–	124	39	3.5
Charleston, W. Va.	43–70	37	14	1.2
Chicago	56–125	63	29	2.9
Cincinnati	47–87	60	24	1.0
Cleveland	58–155	89[c]	26	2.0
Dallas	43–73	59[c]	39	1.4
Denver	80–194	89	37	4.6
Detroit	52–135	68	14	1.8
Houston	51–147	90[c]	39[d]	1.0
Indianapolis	48–81	91[c]	33	2.7
Los Angeles	90	85	117	2.6
Louisville	60–102	70[c]	31	1.5
Milwaukee	47–105	86[c]	41	1.4
Minneapolis	45–87	65[c]	–	1.8
Nashville	41–82	62[c]	49[d]	2.6
New York	40–77	57	35	5.5
Philadelphia	51–109	85	39	3.2
Pittsburgh	88–162	–	29[d]	3.9
St. Louis	63–107	90[d]	22[d]	2.3[d]
San Diego	57–75	69	39	1.1
San Francisco	51	46	20[e]	2.1
Washington, D.C.	47–70	52	29	1.6

[a]EPA (1980).
[b]Annual geometric mean of 24-hour averages.
[c]24-hour averages.
[d]Not a full year.
[e]Total oxidants.

Table 3.2 Ambient Asbestos Levels in Various Cities[a]

	Asbestos Concentration	
Sample Site	ng/m³	Equivalent[b] (fibers/m³) × 10⁻⁴
United States (Selikoff et al., 1972)		
New York City	25–60	1.0–2.4
Manhattan	25–28	1.0–1.1
Bronx	19–22	0.8–0.9
Queens	18–29	0.7–1.2
Staten Island	11–21	0.4–0.8
Philadelphia	45–100	1.8–4.0
Ridgewood, N.J.	20	0.8
Port Allegheny, Penn.	10–30	0.4–1.2
Mean values for 46 of 49 American Cities (NIOSH, 1976)	⩽6	⩽0.2
Maximum mean (Dayton, Ohio)	24	1.0
England (Richards, 1973)		
Rochdale (factory grounds)	1–10	0.04–0.4
Rochdale (town center)	10	0.4
Lancashire/Yorkshire	1–10	0.04–0.4
Industrial Site (Oldbury)	10	0.4

[a]NIOSH (1976).
[b]From Dement et al., 1976, 1 ng asbestos by electron microscopy $\cong$ 400 fibers > 5 μm by phase contrast microscopy.

Table 3.3 Formaldehyde Concentrations Based on Continuous Measurements at Four New Jersey Sites (May 1–September 30, 1974)[a]

Site	Median Daily Average (ppb)	Median Daily 1-hour Maxima (ppb)
Bayonne	6.1	10
Camden	3.8	7
Elizabeth	5.5	10
Newark	6.6	13

[a]Cleveland et al. (1977).

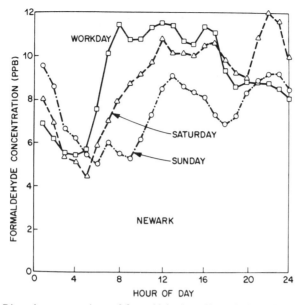

Figure 3.1. Diurnal concentrations of formaldehyde at Newark, New Jersey, for different days of the week, from June 1 to August 31 for the years 1972, 1973, and 1974 (Cleveland et al., 1977).

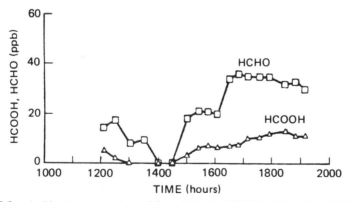

Figure 3.2. Ambient concentrations of formaldehyde (HCHO) and formic acid (HCOOH) as a function of time measured in Riverside, California, October 14, 1977 (Tuazon et al., 1980).

Table 3.4   Ambient Concentration of Some Potentially Hazardous Organic Chemicals[a]

	Two-Week Mean Concentrations (ppb)		
	Los Angeles April 1979	Phoenix April–May 1979	Oakland May–June 1979
Halomethanes			
Methyl chloride	3.00	2.39	1.07
Methyl bromide	0.24	0.07	0.05
Methylene chloride	3.75	0.89	0.42
Chloroform	0.09	0.11	0.03
Carbon tetrachloride	0.22	0.28	0.17
Haloethanes			
1,2 Dichloroethane	0.52	0.22	0.08
1,2 Dibromoethane	0.03	0.04	0.02
1,1,1 Trichloroethane	1.03	0.82	0.29
Chloroethylenes			
Vinyledene chloride	0.005	0.03	0.01
Trichloroethylene	0.40	0.48	0.19
Tetrachloroethylene	1.48	0.99	0.31
Aromatic hydrocarbons			
Benzene	6.04	4.74	1.55
Toluene	11.72	8.63	3.11
Ethylbenzene	2.25	2.00	0.60
m/p - Xylene	4.61	4.20	1.51
o - Xylene	1.93	1.78	0.77
4-Ethyl toluene	1.51	1.51	0.66
1,2,4 Trimethylbenzene	1.88	1.74	–
1,3,5 Trimethylbenzene	0.38	0.40	–
Secondary organics			
Peroxyacetyl nitrate (PAN)	4.98	0.78	0.36
Peroxypropionyl nitrate (PPN)	0.72	0.09	0.15

[a]Singh et al. (1981).

REFERENCES

Cleveland, W. S., Graedel, T. E., and Kleiner, B. (1977). Urban formaldehyde: Observed correlation with source emissions and photochemistry. *Atmos. Environ.* 11:357–360.

Dement, J. M., Zumwalde, R. D., and Wallingford, K. M. (1976). Discussion paper: Asbestos fiber exposures in a hard rock gold mine. *Ann. N.Y. Acad. Sci.* 271:345–352.

EPA (1980). *Air Quality Data–1979 Annual Statistics*. Environmental Protection Agency, Report No. EPA-450/4-80-014, September.

NCRP (1975). *Natural background radiation in the United States.* National Council on Radiation Protection and Measurements Report No. 45, Washington, D.C.

NIOSH (1976). *Revised recommended asbestos standard.* National Institute for Occupational Safety and Health, Report No. 77-169, Washington, D.C., December.

Richards, A. L. (1973). Estimation of submicrogram quantities of chrysotile asbestos by electron microscopy. *Anal. Chem.* 45:809-811.

Scheff, P. A., Holden, J. A., and Wadden, R. A. (1981). Characterization of air pollutants from an activated sludge process. *J. Water Pollut. Control Fed.* 53:223-231.

Selikoff, I. J., Nicholson, W. J., and Langer, A. M. (1972). Asbestos air pollution. *Arch. Environ. Health* 25:1-13.

Singh, H. B., Salas, L. J., Smith, A. J., and Shigeishi, H. (1981). Measurements of some potentially hazardous organic chemicals in urban environments. *Atmos. Environ.* 15:601-612.

Tuazon, E. C., Winer, A. M., Graham, R. A., and Pitts, J. E., Jr. (1980). Atmospheric measurements of trace pollutants by kilometer-pathlength FT-IR spectroscopy. In: *Advances in Environmental Science and Technology,* J. E. Pitts, R. L. Metcalf, and D. Grosjean, Eds., Vol. 10, pp. 259-300, Wiley, New York.

Windham, S. T., Savage, E. D., and Phillips, C. R. (1978). *The effects of home ventilation systems on indoor radon–radon daughter levels.* Environmental Protection Agency, Report No. EPA-520/5-77-011, October.

4

INDOOR SOURCES

The strength and nature of indoor emissions are frequently the most significant consideration in determining indoor air quality. This is obviously true in industrial–occupational settings but is only gradually becoming apparent in nonindustrial situations. The intent of this chapter is to summarize emission data for recognized indoor sources. However, it should be realized that many indoor discharges are not as yet adequately characterized, and that any such summary will only be a beginning.

COMBUSTION

The major combustion sources in American homes are gas-fired appliances. Typically, range, oven, and pilot light emissions are not vented and can contribute to indoor levels of CO, NO, NO_2, and HCHO. Table 4.1 is a compilation of gas-stove emissions. Table 4.2 summarizes emissions from other types of gas-fired appliances but, except for unvented space heaters, most emissions would ordinarily be discharged directly outdoors.

Elevated levels of CO exceeding the 1-hr environmental standard of 35 ppm have been measured in homes with gas stoves (e.g., Sterling and Sterling, 1979; Wade et al., 1975; Hollowell et al., 1977). Nitric oxide and NO_2 levels well above ambient levels have been reported for gas cookers (Melia et al., 1978; Hollowell et al., 1977; Spengler et al., 1979; Wade et al., 1975; Palmes et al., 1977), unvented space heaters (Palmes et al., 1979), and for a forced-air gas-fired central heating system (Hollowell et al., 1977). Figure 4.1 shows mean indoor–outdoor differences for homes with vented and unvented gas and electric stoves (Spengler et al., 1979). Negative values indicate indoor levels lower than outdoor concentrations. All of the values are statistically different from zero ($p < 0.05$). There are no differences among electric cooking values but all gas heating mean concentrations are different from each other ($p < 0.05$). Figure 4.2 illustrates typical increases in CO, NO, and NO_2 when a gas stove is put into use. In Figure

Table 4.1 Emissions from Gas Stoves

| | Heat Input Rate | Emission Factors (μg/kcal) | | | | | | | | | |
| | | Gases, Mean and Standard Deviation | | | | | | | Particulate Matter | | |
		CO	NO	NO_2	SO_2	HCN	HCHO	CO_2	Carbon	Sulfur as SO_4^{2-}	Total Respirable Mass
Range Top Burners	*kcal/burner-hr*										
18 Ranges, 72 burners operated with water-filled cooking pots (Belles et al., 1975; Himmel and Dewerth, 1974)	2270–3020	146[a]	87.7±18.8	36.5±9.7				195,000			
Older stove, cast iron burners											
1 Burner, high flame	2700	382	92.6	51.8							
3 Burners, high flame	2260	475	117.	72.8							
New stove, pressed-steel burners											
1 Burner, high flame	3500	510	130	79.0							
3 Burners, high flame (Yocum, 1981)	3400	315	138	65.6							
4 stoves operated with water-filled cooking pots (Hollowell et al., 1976)	2500	890	31	85	0.8	0.07[b]	5.2	205,000	0.9	0.05	1.7
Japanese stoves											
4 Town gas (4493 kcal/Nm³)	2050			8.7[c]							
2 LPG (23,800 kcal/Nm³) (Yamanaka et al., 1979)	3100			7.9[c]							
American Gas Association Standard		645									
British Gas Corporation	2500			136[c]							
Pilot Lights	*kcal/hr*										
9 Range top (average)	40.1–43.8	140[d]	39.9±3.2	23.4±3.2				195,000			
3 Free-standing single flame with flashtube		114	43.6	24.9							
3 Free-standing single flame with flashtube and shield around flame		88	41.3	21.3							

Table 4.1 *(Continued)*

	Heat Input Rate	Emission Factors ($\mu g/kcal$)									
		Gases, Mean and Standard Deviation							Particulate Matter		
		CO	NO	NO_2	SO_2	HCN	HCHO	CO_2	Carbon	Sulfur as SO_4^{2-}	Total Respirable Mass
Pilot Lights	*kcal/hr (Continued)*										
3 Free-standing single flame with flashtube and shield around and above flame (most popular)		182	38.6	23.6				195,000			
5 Range ovens and broiler (average)	43.1–44.6	1043[e]	2.3±3.6	54.1±10.6							
3 Constant-input pilot		1376	1.1	61.3							
2 Standby pilot mode (tested) and separate ignition mode		829	3.9	42.6							
(Belles et al., 1975; Himmel and Dewerth, 1974)											
Range top											
Cast iron burners	150	419	45.3	54.6							
Pressed-steel burners	100	842	4.7	18.6							
(Yocum, 1981)											
Ovens and Broilers	*kcal/hr*										
27 Units	2770–6050	105[f]	97.7±24.0	31.0±10.7				195,000			
(Belles et al., 1975; Himmel and Dewerth, 1974)											
Steady state											
Older oven	2200	530	91.4	73.1							
Newer oven	2200	1620	77.9	50.4							
(Yocum, 1981)											
Steady state											
11 ovens at 180°C (350°F)	2000	950	29	62	0.8	1.8[g]	11.4	200,000	0.13	<0.01	
(Hollowell et al., 1976)											
American Gas Association Standard		645		85[c]							
British Gas Corporation											

[a] Two-thirds of the data included within 35% and 284% of the average.
[b] One run.
[c] NO + NO_2.
[d] Two-thirds of the data included within 69% and 145% of the average.
[e] Two-thirds of the data included within 59% and 169% of the average.
[f] Two-thirds of the data included within 40% and 247% of the average.
[g] Average of three runs.

Table 4.2   Emissions from Gas-Fired Appliances

Appliances	Heat Input Rate (kcal/hr)	Emission Factors (μg/kcal), Mean and Standard Deviation		
		CO	NO	NO$_2$
38 Forced-air furnaces[a,b,c]	18,900–45,200	16 ± 24	111 ± 26	7.8 ± 3.9
9 Forced-air furnace pilots	209–396	199 ± 73.9	44 ± 9	55 ± 13.8
Water Heaters				
United States[a,d]				
17 Domestic (20–50 gal capacity)	6930–12,600	8 ± 3.9	135 ± 37	9.0 ± 3.8
3 Commercial (multiport) (85 gal)	62,000–96,000	8	221	10
1 Commercial (single port) (100 gal)	48,000	5	93.9	7.2
13 Pilot lights	175–312	278 ± 165	37.6 ± 19	57.5 ± 17
Japan[e]				
4 Domestic	8770		129[f]	
Space Heaters				
United States				
9 Vented[a]		8	165	11.9
6 Unvented[a]		20	133	24.3
1 Unvented[g]	2800 (low flame)	632	76.4	46.4
	6200 (high flame)	319	135	43.8
8 Pilot lights[a]	85–256	139	54.4	46.1
Japan[e]				
6 Radiant unvented	2010[h]		46[f]	
5 Convective unvented	2110[h]		251[f]	
Hot-Water Boilers[a]				
4 Domestic multiport		7	194	21.9
1 Domestic single port		9	80.8	5.7
2 Swimming pool heaters		61	298	48.1

[a]Belles et al., 1975; natural gas-fired ($\sim$9250 kcal/N m^3).
[b]Thrasher and Dewerth, 1975.
[c]Also measured 0.23 ± 0.18 μg/kcal of total aldehydes expressed as HCHO.
[d]Thrasher and Dewerth, 1977.
[e]Yamanaka et al., 1979; town gas fired ($\sim$4990 kcal/N m^3).
[f]NO + NO$_2$.
[g]NAS, 1981b.
[h]Kerosene fuel ($\sim$10,990 kcal/kg).

4.2B, which is for a house with an electric stove, there appears to be a small O$_3$ contribution from the stove and also a significant NO$_x$ increase with use of a forced-air heating system (Hollowell et al., 1977). The small amount of sulfate from gas ovens reported in Table 4.1 (possibly produced from mercaptans added

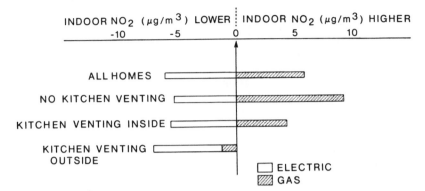

Figure 4.1. Mean indoor–outdoor difference in nitrogen dioxide concentrations by cooking fuel and kitchen ventilation; average across all indoor–outdoor sites (May 1977–April 1978) [Spengler et al. (1979). Reprinted by permission of The American Chemical Society from *Environmental Science and Technology*.]

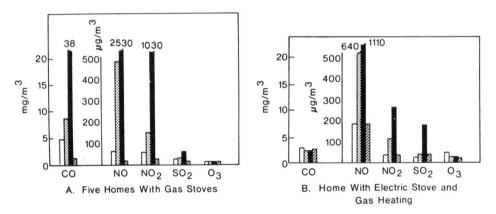

Figure 4.2. Gaseous indoor and outdoor air pollutant levels observed at houses with various cooking and heating configurations (Hollowell et al., 1977).

A: ☐ kitchen—mean value with pilot light only

     ▦ kitchen—mean value during cooking

     ■ kitchen—maximum value during cooking

     ▨ outdoor

B: ☐ kitchen—electric stove ON, gas heating OFF

     ▦ bedroom—electric stove OFF, gas heating ON intermittently for approximately 8 hr

     ■ bedroom—electric stove OFF, gas heating ON continuously for 50 min.

     ▨ outdoor

Table 4.3   Radon Concentration in Natural Gas in the Distribution Line[a]

Area	Radon Concentration (pCi/liter)	
	Average	Range
Poland (Warsaw)	8	4–14
United States		
Chicago	14.4	2.3–31.3
New York City	1.5	0.5–3.8
Denver	50.5	1.2–119
West coast	15	1–100
Colorado	25	6.5–43
Nevada	8	5.8–10.4
New Mexico	45	10–53
Houston	8	1.4–14.3

[a]UN (1977).

Table 4.4   Emissions of Some Pollutants from Wood Burning Stoves and Fireplaces[a]

Pollutant	Emissions, g/kg Wood (Range)	
	Stoves	Fireplaces
CO	160(83–270)	22(11–40)
NO_x as NO_2	0.5(0.2–0.8)	1.8
SO_x as SO_2	0.2(0.15–0.45)	—
Formaldehyde	0.2(0.1–0.3)	0.4
Acetaldehyde	0.1	—
Phenols	1.0(0.1–2.4)	0.02
Total particulate matter	9.2(1–28)	9.1(7.2–12)
Benzo(α)pyrene	0.0025	0.00073

[a]After Cooper (1980); Duncan et al. (1980); Butcher and Buckley (1977); DeAngeles et al. (1979).

to the gas as odiferants) was also supported by some weak statistical evidence from a six-city (United States) study on indoor-outdoor air pollution relationships (Dockery and Spengler, 1981b).

The natural gas used for cooking and heating may also be a source of Rn-222. However, as Table 4.3 indicates, the amount of radiation released will probably not be excessive.

Another combustion source which is of emerging interest is the wood stove. Since combustion is much less complete than for oil and gas furnaces, emissions will be greater. Table 4.4 summarizes some of the data on wood stoves and fireplaces. Although most of the emissions are expected to be vented outside, leaks and improper operation of flue systems may cause indoor emissions. For certain types of stoves, emissions were found to increase substantially with reduction in draft setting (Butcher and Buckley, 1977). Indoor measurements in Boston area homes with wood stoves indicated that average indoor TSP and benzo(α)pyrene levels during wood burning were 3 and 5 times, respectively, the levels when wood was not burned (Zabransky, 1981).

SMOKING

Tobacco smoke contains a great variety of potentially hazardous materials. Actual emission factors per cigarette for a number of these substances are given in Table 4.5. Mainstream smoke is that which is inhaled by a smoker and the emission factors are representative of preinhalation conditions. Many of the pollutants will be filtered out in the smoker's lungs, e.g., 70% of the particulate matter (Hoegg, 1972). Sidestream smoke is primarily the unfiltered smoke emitted from an idling cigarette, cigar, or pipe. The emission factors for sidestream smoke are consequently more useful for characterization of indoor environments where smoking is allowed.

Actual measurements in environmental settings have been given in Table 2.6. In addition, in a recent study of respirable particles inside and outside homes in six American cities, smoking was found to be the major source of indoor particulate matter. Each smoker in a home was shown to contribute approximately 20 $\mu g/m^3$ to the annual indoor respirable particle concentration (Dockery and Spengler, 1981a, b; Spengler et al., 1981). Approximately 50-75% of American homes have been reported to have one or more smokers (NAS, 1981b).

The particle sizes shown in Table 4.5 indicate that smoke particles are likely to penetrate and deposit in the alveolar spaces of the lung. More recent mainstream data indicate initial mass median diameters of 0.37-0.52 μm with growth to 0.42-0.71 μm within 6-15 sec (Hinds, 1978). Sidestream particles are also persistent. Seventy-five percent remained suspended in a test chamber after 2.5 hr with a median size of 0.7 μm and no particles >2 μm (Hoegg, 1972).

Table 4.5 Emission Factors for Mainstream and Sidestream Smoke

Properties	Source[a]	Mainstream	Sidestream	Sidestream/ Mainstream Ratio
General Characteristics				
Duration of smoke production	2	20 sec	550 sec	27
Amount of tobacco burnt	2	347 mg	411 mg	1.2
Number of particles per cigarette	2	1.05×10^{12}	3.5×10^{12}	3.3
Particle number median diameter	2	0.2 μm	0.15 μm	0.75
Particulate Phase		μg/cigarette	μg/cigarette	
Total suspended particulate matter	2	36,200	25,800	0.7
Tar (chloroform extract)	9, 2	<500–29,000	44,100	2.1
Nicotine	4, 9	100–2500	2700–6750	2.7
Total phenols	2	228	603	2.6
Pyrene	4	50–200	180–420	3.6
Benzo(α)pyrene	4	20–40	68–136	3.4
Naphthalene	4	2.8	40	16
Methylnaphthalene	4	2.2	60	28
Aniline	4	0.36	10.8	30
NNN[b]	4	0.1–0.55	0.5–2.5	5
NNK[c]	4	0.08–0.22	0.8–2.2	10
Cadmium	2	0.13	0.45	3.6
Nickel	3	0.08	–	–
Arsenic	3	0.012	–	–
2-Naphthylamine	4	0.002–0.028	0.08	39
Hydrogen cyanide	5	74	–	–
Polonium-210	7	0.029–0.044 pCi/cigarette	–	–
Gases and Vapors		μg/cigarette	μg/cigarette	
Carbon monoxide	4, 9	1000–20,000	25,000–50,000	2.5
Carbon dioxide	4	20,000–60,000	160,000–480,000	8.1
Acetaldehyde	1, 3	18–1400	40–3100	2.2[d]
Hydrogen cyanide	4	430	110	0.25
Methylchloride	4	650	1,300	2.1
Acetone	4	100–600	250–1500	2.5
Ammonia	1, 3, 4	10–150	980–150,000	98
Pyridine	1, 3, 4	9–93	90–930	10
Acrolein	1, 3	25–140	55–300	2.2[d]
Nitric oxide	3, 6, 8	10–570	2300[e]	4
Nitrogen dioxide	3, 6, 8	0.5–30	625[e]	20
Formaldehyde	3, 6	20–90	1300	15
DMN[f]	4	10–65	520–3380	52
NPy[g]	4	10–35	270–945	27

[a]*Sources:* (1) Wakeham (1972); (2) Hoegg (1972); (3) HEW 1979, Chap. 14; (4) HEW (1979), Chap. 11; (5) Schmeltz et al. (1975); (6) Weber et al. (1979); (7) Kelley (1965); (8) Jenkins and Gill (1980); (9) FTC (1981).

[b]Nitrosonornicotine (NNN).

[c]4-(*N*-methyl-*N*-nitrosamino)-1-(3-pyridyl)-1-butanone (NNK).

[d]Ratioed from total aldehydes.

[e]Based on the data of Weber et al. (1979), Table 1, assuming a mixing factor of 1.0. Values will be lower if mixing was less than ideal.

[f]Dimethylnitrosamine (DMN).

[g]Nitrosopyrolidine (NPy).

BUILDING MATERIALS

Radon Daughters

Various types of construction materials have been identified as sources of hazardous materials. Table 4.6 shows typical Ra-226 levels of rock, sand, and clay products. Ra-226 is the source of gaseous Rn-222 emissions. Table 4.7 indicates the Rn-222 emanation rate per unit activity concentration of Ra-226 ($pCi/m^2 \cdot s$ per pCi/g). The diffusion of Rn-222 from building materials is influenced by moisture content of the material, density, the presence of sealants, the nature of the material itself, and the nature of the substances with which it is mixed. The fractional escape (pCi escaped per pCi produced), another measure of Rn-222 release, is typically of the order of 1% for building materials in walls and ceilings. Both emanation rates and fractional escape are functions of material thickness (UN, 1977). Radon can also diffuse through earth, basement walls, and floors in contact with soil containing radioactive materials. The average Rn-222 emanation rate for soil is 0.42 $pCi/m^2 \cdot s$ with a range of 6×10^{-3}–1.4 $pCi/m^2 \cdot s$ (UN, 1977). Since the activity of both Ra-222 and its short half-life daughters are of interest, it is necessary to solve a set of five differential equations which describe the decay and venting of radon daughters to Pb-210 (which terminates the series because of its 22-year half-life). These can be described by

$$\frac{dA_0}{dt} = Q - (\lambda_0 + I) A_0 + IA_0^o. \tag{4.1}$$

and for $i = 1, 2, 3$, or 4,

$$\frac{dA_i}{dt} = \lambda_i A_{i-1} - (\lambda_i + I) A_i + IA_i^o \tag{4.2}$$

where subscript $i = 0$ through 4 denotes Rn-222, Po-218, Pb-214, Bi-214, and Po-214, respectively; A_i is the activity concentration of the ith radionuclide (pCi/liter) at time t; λ_i is its decay constant (min^{-1}); the absence of a superscript on A_i refers to indoor values while superscript o refers to outdoor values; I is the air exchange rate (air changes/min); Q is the source strength, the entry rate of Rn-222 into the building per unit volume (pCi/liter-min) (Kusuda et al., 1980). Table 4.8 from the same reference gives appropriate constants. Working level (WL) is a measure of total potential energy concentration (E) of the daughters. One WL is defined as any combination of short-lived radon daughters ultimately yielding 1.3×10^5 MeV of potential energy per liter of air upon decay

Table 4.6   Concentration of Ra-226 in Building Materials[a]

Type of Building Material	Country	Number of Samples	Average Activity Concentration Ra-226 (pCi/g)
Bricks	F.R. Germany	132	2.6
Bricks	Sweden	21	2.6
Red bricks	U.S.S.R.	55	1.5
Clay bricks	U. Kingdom	23	1.4
Red-slime bricks	F.R. Germany	23	7.6
Concrete	F.R. Germany	69	1.8
Heavy concrete	Sweden	15	1.3
Aerated concrete without alum shale	Sweden	22	1.5
Concrete containing alum shale	Sweden	83	40.4
Heavy concrete	U.S.S.R.	87	0.9
Light concrete	U.S.S.R.	16	2.0
Concrete	U. Kingdom	5	2.0
Cement	F.R. Kingdom	19	1.2
Cement	Sweden	8	1.5
Cement	U.S.S.R.	7	1.2
Natural plaster	F.R. Germany	23	< 0.5
Natural plaster	Sweden	4	0.09
Plaster	U.S.S.R.	1	0.25
Natural plaster	U. Kingdom	69	0.6
Phosphogypsum			
From apatite	F.R. Germany	2	1.5
From phosphorite	F.R. Germany	39	16
Of unknown origin	F.R. Germany	7	< 0.5
Phosphogypsum	U. Kingdom	6	21
Phosphogypsum	United States	–	40
Granite	F.R. Germany	34	2.6
Granite	U.S.S.R.	2	3
Granite bricks	U. Kingdom	7	2.4
Pumice stone	F.R. Germany	20	3.0
Tuff	U.S.S.R.	13	2.6
Limestone and marble	F.R. Germany	20	< 0.5
Rock aggregate	Sweden	296	1.3
Rock aggregate	U. Kingdom	3	1.4
Gravel and sand	F.R. Germany	50	< 0.4
Natural sand and sand rejects	U.S.S.R.	32	0.63
Blast-furnace slags	U.S.S.R.	29	1.8
Wood	Sweden	1	–
Rock and silica wool	Sweden	2	0.4
Rock and silica wool	U. Kingdom	2	Negligible
Lightweight aggregate	Sweden	10	3.9
Fly ash	F.R. Germany	28	5.7
Fly ash (mixture of coal clinker, ash and cement)	U. Kingdom	3	0.2–3.7

[a] UN (1977).

Table 4.7   Radon Emanation Rates of Various Materials[a]

Material	Emanation Rate of Rn-222 per Unit Activity Concentration of Ra-226 ($pCi/m^2 \cdot s$ per pCi/g)	Comments
By-product gypsum	0.01	Internal walls 76 mm thick
By-product gypsum	0.001	Ceilings 13 mm thick
Concrete	0.005	10 cm thick
Uranium mill tailings	0.2	10 cm thick
Uranium mill tailings	1.6	"Infinite" thickness
Soil	0.5	"Infinite" thickness
Light concrete	0.02	20 cm thick
Heavy concrete	0.01	8 cm thick

[a]UN (1977).

to Pb-210. [It is further estimated that 1 WL-month (WLM) will deliver about 1 rad of alpha radiation to the bronchial epithelium, the probable site of lung cancer incidence from these materials (UN, 1977).] In general,

$$E = \sum_{i=1}^{4} C_i A_i \qquad (4.3)$$

where the C_i values are given in Table 4.8. The unsteady-state solution of Equations (4.1) and (4.2) as well as working level values have been calculated by Kusuda et al. (1980). Figures 4.3 and 4.4 show typical concentration patterns for various indoor source strengths and ventilation rates.

Formaldehyde

Particle board and urea-formaldehyde foam insulation have been identified as HCHO emission sources (Andersen et al., 1975; Allan et al., 1980; Sardinas et al., 1979; Breysse, 1981; Dally et al., 1981). HCHO concentrations of 60–1673 ppb (0°C, 1 atm) with an average of 463 ppb were measured in 25 rooms in 23 Danish dwellings where chip board was used in walls, floors, and ceilings (Andersen et al., 1975). An empirical mathematical model was developed for room air concentration:

$$C = \frac{(RT - N)(aH + b)}{1 + nf/\alpha} \qquad (4.4)$$

where C is air concentration in mg HCHO/m^3; $R = 0.064$, $a = 0.143$, $b = 0.048$,

Table 4.8 Potential Alpha Energy of Rn-222 and Its Short-Lived Decay Products[a]

Radionuclide	i	Radioactive Half-Life	Radioactive Decay Constant λ_i (min^{-1})	Number of Atoms/Picocurie	Potential Alpha Energy (MeV) per Atom	per Picocurie	Conversion Factor C_i $\dfrac{\text{MeV/liter}}{\text{pCi/liter}}$	$\dfrac{\text{WL}}{\text{pCi/liter}}$
Ra-226 $\downarrow$								
Rn-222	0	3.8d	1.258×10^{-4}	17,488				
Po-218(RaA)	1	3.05 min	0.2272	9.77	13.68	134	134	0.00103
Pb-214(RaB)	2	26.8 min	0.02586	85.3	7.68	659	659	0.00507
Bi-214(RaC)	3	19.7 min	0.03518	63.1	7.68	485	485	0.00373
Po-214(RaC')	4	1.6×10^{-4} s	2.60×10^5	10^{-5}	7.68	7.68×10^{-5}	7.68×10^{-5}	6×10^{-10}
Pb-210 $\downarrow$								

[a]Kusuda et al. (1980).

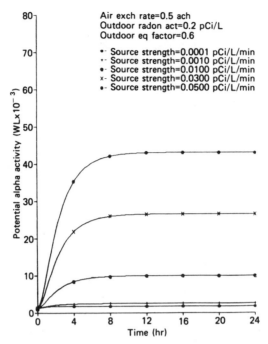

Figure 4.3. Potential alpha activity (E) for various source strengths; $A_{IO} = A_i^O$, $I = 0$–4 (Kusuda et al., 1980).

$f = 0.304$, and $N = 0.764$ are empirical constants from the testing program; H is humidity, g H_2O/kg dry air; T is temperature, °C; n is air change rate per hour; α is area of board surface per volume of room, m^{-1}. The equation was found to be applicable for the following ranges: 17–32°C, 5–13 g H_2O/kg air, and 0.4–3 air changes/hr. Since the concentration data supporting Equation (4.4) were all taken after equilibrium had been reached between emissions and removal by ventilatory flow, formaldehyde release from particle board can alternatively be represented by an emission factor S', where

$$S' = \frac{nC_{i,ss}}{\alpha} = \frac{(kn/\alpha)(RT - N)(aH + b)}{1 + nf/\alpha} \qquad (4.5)$$

where $C_{i,ss}$ is the steady-state indoor concentration and k is a term to describe the efficiency of mixing (see Chapter 6). For the factors defined as above, S' will have the units of mg HCHO/hr · m^2 of board surface. Using average parameter values of $n/\alpha = 0.67$ m/hr, $T = 22.8$°C, and $H = 7.1$ g/kg (from Andersen et al., 1975) and assuming perfect mixing ($k = 1$) results in a value of $S' = 0.41$ mg/m^2 · hr. This value is somewhat larger than those derived from the studies of Berge and Mellegaard (1979) in a closed system where S' was approximately 0.035

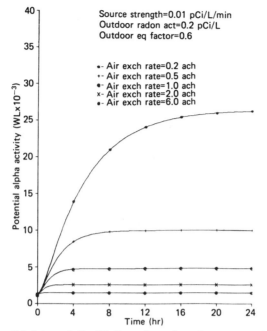

Figure 4.4. Potential alpha activity (E) for various air exchange rates; $A_{IO} = A_i^O$, $I = 0-4$ (Kusuda et al., 1980).

mg/m² · hr at 22-25°C at an undetermined relative humidity. This range of variation is perhaps not unexpected for particle board emissions. In addition, more realistic values of k range between 0.1 and 0.3 which would bring the two values closer together (see *Mixing Factors* in Chapter 6). Also of interest is the observation that HCHO emissions from particle board aged for 13 months were still about 70% of originally measured values (Andersen et al., 1975).

A survey of 334 mobile homes (608 samples) in the state of Washington revealed HCHO concentrations from 40 μg/m³ (30 ppb) to 2370 μg/m³ (1770 ppb), 66% between 130 and 660 μg/m³ (100-490 ppb), and 21% > 670 μg/m³ (500 ppb). A total of 523 persons experienced one or more symptoms which could reasonably be related to HCHO exposure. Particle board and plywood were identified as the major HCHO sources (Breysse, 1981).

There is some evidence, based on laboratory and home measurements, that emissions from particle board decrease exponentially with time. The relationship between indoor HCHO concentration and time since installation, t, can be expressed as

$$C = G_1 e^{-G_2 t} \tag{4.6}$$

where G_1 and G_2 are empirical constants. The equation does not take into account amount of particle board, temperature, humidity, or ventilation. The half-life based on Equation (4.6) has been variously reported as 2 years (Scandinavian home construction particle board), 58 months (Danish homes), 69 months (randomly selected Wisconsin mobile homes), and 28 months (complained-about Wisconsin mobile homes). When these data are combined, the half-life is 53 months (NAS, 1981b), although the variation from this value for individual homes may be considerable. An analysis of $\frac{1}{2}$-hr samples taken from mobile homes in Minnesota demonstrated the same general trend (Garry et al., 1980). Whether reduction in formaldehyde concentration with time was due to reduced off-gassing or because of a change in the binding agent formulation was not determined.

Urea-formaldehyde foam is produced when the two major constituents are combined with a catalyst and forced from a pressurized nozzle. The product is cured to a hardened resin. The chemical reaction is

$$\underset{H_2N\overset{O}{\overset{\|}{C}}NH_2}{} + HCHO \rightleftharpoons \underset{H_2N\overset{O}{\overset{\|}{C}}{-}NH{-}CH_2OH}{} \tag{4.7}$$

The degree to which formaldehyde may be emitted from urea-formaldehyde foam is directly related to the composition of the precursor resin (a function of manufacturing conditions) and the extent of breakdown reactions in the foamed product (which depends on the acidity of the foam *in situ*) (Allan et al., 1980). Although improper curing is usually cited as being the cause of HCHO release, it is unclear whether this takes place during manufacture, mixing, or installation (Sardinas et al., 1979).

Emission factors for several commercial UF foams are given in Table 4.9. The emission data (Long et al., 1979; Allan et al., 1980) were collected under laboratory conditions from foams which met suggested standards (Rossiter et al., 1977) and were prepared with commercial equipment. The emission factors are based only on measurements made 9 or 10 days after foaming. Emission factors at shorter times are higher (by perhaps factors of 2–4) but their calculation is complicated by rapid concurrent changes in foam density, particularly at high humidities. It is suggested that the >10-day averages are more appropriate for design evaluation with the recognition that at much longer times the emission factor may decrease further. Although emissions from the various foams did not differ greatly, it has been pointed out that extractable HCHO, which may reflect total potential discharge over the life of the foam, may differ by a factor of 10 between different formulations (Long et al., 1979).

Levels greater than 500 $\mu g/m^3$ (374 ppb) (the minimum detectable limit of the sampling procedure) up to 10,000 $\mu g/m^3$ (7400 ppb) were measured for 38 out of 68 Connecticut homes (Sardinas et al., 1979). All of these householders had complained of HCHO exposure centering on insulation which had been installed from 3 weeks to 1.5 years before testing. Following foam insulation in 39

Table 4.9   Emission Factors for Several Urea-Formaldehyde Foams

	Number of Observations	Emission Factor (μg HCHO/g foam · min)	
		Mean	Standard Deviation
Average of data from 3 commercial foams tested at 33°C under 10 and 85% relative humidities over the interval of 10–30 days after foaming. (Based on data from Long et al., 1979)	42	0.109	0.050
85% Relative humidity only	21	0.122	0.040
10% Relative humidity only	21	0.095	0.056
Commercial foam exposed at 35°C and 90% relative humidity over the interval of 9–26 days. (Based on data from Allan et al., 1980)	2	0.170	

Table 4.10   Indoor–Outdoor Formaldehyde and Aliphatic Aldehyde Concentrations Measured at the Med-II Residence, August 1979[a]

Condition	Number of Measurements	Sampling Time, hr	Formaldehyde (ppb)[b]	Aliphatic Aldehydes (ppb)[c]
Unoccupied, without furniture	3	12	66 ± 9%	74 ± 16%
Unoccupied, with furniture	3	24	183 ± 7%	241 ± 4%
Occupied, day[d]	9	12	214 ± 10%	227 ± 15%
Occupied, night[e]	9	12	115 ± 31%	146 ± 29%

[a]Chin-I-Lin et al. (1979).
[b]Determined using pararosaniline method (100 ppb $\cong$ 120 μg/m^3). All outside concentrations < 10 ppb.
[c]Determined using MBTH method, expressed as equivalent of formaldehyde. All outside concentrations < 20 ppb.
[d]Air exchange rate/hr $\cong$ 0.4.
[e]Windows open part of time; air exchange rate significantly greater than 0.4 ach and variable.

Washington State homes, 73% of 93 HCHO samples were > 130 μg/m^3 (100 ppb) and 20% > 670 μg/m^3 (500 ppb). Forty-four persons experienced symptoms which could be associated with HCHO exposure (Breysse, 1979).[*]

[*]In February 1982, the U.S. Consumer Product Safety Commission voted to ban the use of urea formaldehyde insulation in schools and homes (but not hotels, factories, office buildings, and other similar buildings), and Canada banned further installations in December 1980 (C&EN, 1982).

Formaldehyde concentrations have also been measured in an "energy tight" house with and without furniture. As Table 4.10 shows, furniture adhesives, plywood, and textiles may also constitute a HCHO source (Chin-I-Lin et al., 1979).

Asbestos

Asbestos fiber contamination of a building interior occurs by fallout, contact or impact, and reentrainment (Sawyer and Spooner, 1978). The rate of fiber release in fallout is continuous, low-level, and persistent. Fallout may occur without actual physical disruption of the fiber-bearing material and may be a function of the degradation of the adhesive binder. The rate may vary due to structural vibrations, humidity changes, and air movements. The range of concentrations is roughly from nearly zero for cementitious mixes in good repair to 100 ng/m^3 for deteriorating dry mix applications. Concentrations measured during periods of quiet activity have varied from near ambient to approximately 100 ng/m^3 by electron microscopy and 2×10^4 fibers/m^3 by optical microscopy (Sawyer and Spooner, 1978).

Table 4.11   Airborne Asbestos in Buildings[a]

Sampling Conditions or Situation	Mean Counts (fibers/cm^3)	Number of Samples	Standard Deviation
OSHA 8-hr standard	2		Maximum excursion to 10
NIOSH 1977 recommended standard	0.1		Maximum excursion to 0.5
University dormitory, UCLA Exposed friable surfaces, 98% amosite General student activities	0.1	—	0–0.8 (range)
Art and Architecture Building, Yale University. Exposed friable ceilings, 20% chrysotile Ambient air, City of New Haven	0.00	12	0.00
Fallout			
Quiet conditions	0.02	15	0.02
Contact			
Cleaning, moving books in stack area	15.5	3	6.7
Relamping light fixtures	1.4	2	0.1
Removing ceiling section	17.7	3	8.2
Installing track light	7.7	6	2.9

Table 4.11 *(Continued)*

Sampling Conditions or Situation	Mean Counts (fibers/cm^3)	Number of Samples	Standard Deviation
Installing hanging lights	1.1	5	0.8
Installing partition	3.1	4	1.1
Reentrainment			
Custodians sweeping, dry	1.6	5	0.7
Dusting, dry	4.0	6	1.3
Proximal to cleaning (bystander exposure)	0.3	–	0.3
	0.2	36	0.1
General Activity			
Office buildings, eastern Connecticut			
Exposed friable ceilings, 5 to 30% chrysotile			
Custodial activities, heavy dusting	2.8	8	1.6
Private homes, Connecticut			
Remaining pipe lagging (dry) amosite and chrysotile asbestos	4.1	8	1.8–5.8 (range)
Laundry: contaminated clothing. chrysotile	0.4	12	0.1–1.2 (range)
Office building, Connecticut			
Exposed sprayed ceiling, 18% chrysotile			
Routine activity	79[b]	3	40–110 (range)
Under asbestos ceiling	99[b]	2	
Remote from asbestos ceiling	40[b]	1	
Urban Grammar School, New Haven			
Exposed ceiling, 15% chrysotile asbestos			
Custodial activity: sweeping, vacuuming	643[b]	2	186–1100 (range)
4 Massachusetts Schools (Irving et al., 1980)	0.06–0.16		
10 New York, New Jersey and Massachusetts Schools with damaged asbestos surfaces (Nicholson et al., 1979)	217[b]		9–1950 (range)
Apartment Building: New Jersey, heavy housekeeping. Tremolite and chrysolite	296[b]	1	
Office Buildings, New York City			
Asbetos in ventilation systems	2.5–200[b]		0–800 (range)
Quiet conditions and routine activity			

[a]Sawyer and Spooner (1978). [b]Nanograms/cubic meter. Determined by electron microscope.

Fiber dispersal during routine maintenance may release in excess of 20×10^6 fibers/m^3 (contact mode), and removal of dry sprayed asbestos can yield over 100×10^6 fibers/m^3 (Sawyer, 1977). Reentrainment can cause fiber counts of 5×10^6 fibers/m^3 during custodial work. Emission factors are not available for estimating asbestos release. However, Table 4.11 lists a variety of measurements in different environments not related to the manufacturing or processing of asbestos (Sawyer and Spooner, 1978).

OFFICE MACHINES AND DOMESTIC AIR CLEANERS

Tests on photocopying machines (Selway et al., 1980; Allen et al., 1978) and domestic and commercial size electrostatic air cleaners (Allen et al., 1978; Sutton et al., 1976) have shown these devices to be indoor ozone sources. Appropriate emission factors are given in Table 4.12. For copying machines, operator

Table 4.12   Ozone Emissions from Copying Machines and Domestic Air Cleaners

	Maximum Voltage	Ozone Emission Factors
		μg/min
Electrostatic Air Cleaners		
8 Installed in central air conditioning systems[a]	5000–7900	0–546
"Several well-known manufacturers' electronic air cleaners" (on central air conditioning systems)[b]		303–1212
1 Portable unit[a]	9900	84
Two-stage, low-voltage industrial unit (1 pass; 2 passes will double the emission rate)[c]	11,000	333
		μg/copy[e]
11 Photocopying Machines[a,d]	3500–11,000	Range <2–158; Typically 15–45

[a]Allen et al. (1978).
[b]Sutton et al. (1976).
[c]Holcomb and Scholz (1981).
[d]Selway et al. (1980).
[e]Typical copy rate, 5 copies/min.

breathing-zone concentrations up to equilibrium values of 0.068 ppm were measured under normal working conditions. Copying machine cleaning and maintenance were shown to temporally reduce emission but with return to previous rates within 15 days (Selway et al., 1980). Hydrocarbon discharges (measured as petroleum distillate) have resulted in air concentrations up to 1.05 mg/m^3 over 7.5 hr (NIOSH, 1977), although emission data are not available. There have also

Table 4.13   Concentrations of Organic Vapors in Printmaking Areas and a Plastics Mold Room[a]

Sample Description	Substance	Concentration (mg/m^3)
Etching—BZ[b]	Lithotine/benzine[c]	6.7
Etching—BZ	Lithotine/benzine	50.0
Etching—BZ	Lithotine/benzine	20.8
Photoetching—BZ	Acetone	91.5
	Benzene	112.7
	Methyl cellosolve acetate[d]	176.1
Silkscreening—BZ	Mineral spirits[c]	163.2
	Toluene	10.4
	Xylene	2.6
Silkscreening—BZ	Mineral Spirits	34.1
	Toluene	trace
	Xylene	trace
Etching—BZ	Lithotine/benzine	16.8
Etching—area	Lithotine/benzine	13.8
Plastics molding—BZ	Acetone	25.0
	Toluene	trace
	Styrene	2.9
Plastics molding—area	Acetone	56.2
	Toluene	trace
	Styrene	trace
Plastics molding—area	Acetone	43.3
	Toluene	2.3
	Styrene	10.0

[a]NIOSH (1977).

[b]BZ, breathing zone.

[c]Lithotine®, Benzine®, and Mineral Spirits® are the trade names of the bulk liquids used in these areas.

[d]Ethylene glycol monomethyl ether acetate.

Table 4.14   Emissions of Chrysotile Asbestos from 21
Hand-Held Hair Dryers[a]

	Maximum	Median	Minimum
Emission Rate[b]			
Fibers/hr	1.9×10^6	1.1×10^5	7×10^3
Total mass, ng/hr	3305	220	1

[a]Hallenbeck (1981).
[b]Samples of emissions were analyzed by transmission electron microscopy. Data are derived from 2-hr high-heat test runs only. There was
no statistical difference between results obtained from 2-hr high-heat
test runs and those obtained by cycling between high and low heat for
2 hr. Data under column headings do not necessarily represent tests on
the same hair dryer (Geraci et al., 1979).

been reports that some toners used in photocopying machines contain mutagens,
possibly due to trace amounts of nitropyrene (C&EN, 1980).

OTHER SOURCES

Emissions from industrial strength ammoniated cleaning compounds and germi-
cidal cleaners can release a variety of organic compounds (Selway, 1979; Allen,
1977). Furnishings, adhesives, and paints are also organic sources (Hollowell
et al., 1979) but without well-described emission factors. Hobbies can produce

Table 4.15   Typical Aerosol Size Distributions for Estimating Inhalation Exposure from
Pressurized Consumer Products[a]

Product Type	Total Aerosol Concentration (mg/m^3)	Mass Median Aerodynamic Diameter (μm)	Weight % of Aerosol in Each Aerodynamic Diameter Range		
			$<1\ \mu m$	$1-3\ \mu m$	$3-6\ \mu m$
Air freshener	27	5.2–6.3	5	16	32
Antiperspirant	246	5.9–7.3	3	17	30
Dusting aid	86	6.4–7.5	2	10	30
Fabric protector	9	2.6–4.0	13	30	28
Furniture wax	22	3.0–4.9	11	29	34
Hair spray	30	5.8–6.4	5	16	29
Paint	189	7.2–8.7	2	8	22
Wood panel wax	15	1.4–1.5	2	5	22

[a]Mokler et al. (1979a).

organics, metal fumes (from welding and metal working), and respirable particles (NIOSH, 1976; NIOSH, 1979). Table 4.13 shows the measured concentrations of various chemicals in work areas of an art school (NIOSH, 1976). Sulfur in clay can be released from a kiln as SO_2. Methylene chloride, a paint stripping ingredient, ties up red blood cells in a fashion similar to CO (Stewart and Hake, 1976). Barium poisoning can occur while firing glazes containing barium carbonate.

Domestic activities are another source of hazardous materials. Asbestos is discharged from hair dryers and typical emission data are given in Table 4.14. Air fresheners, furniture waxes, and paints are typical of home products delivered in aerosol form. Table 4.15 gives typical size distributions for a variety of aerosol products dispersed in user breathing zones. For a given product category, the size distributions appeared to be reasonably consistent between brands and for different methods of application (Mokler et al., 1979a, b). However, the authors suggest that the measured concentrations (collected in a 1-m^3 chamber

Table 4.16   Materials Emitted by Humans[a]

	Typical Concentrations (ppb) (389 People in Lecture Class at 9:30 a.m.)	Emission Rate (mg/day per Person)	
		Lecture Class	Exam
Organic Bioeffluent			
Acetone	20.6 ± 2.8	50.7 ± 27.3	86.6 ± 42.1
Acetaldehyde	4.2 ± 2.1	6.2 ± 4.5	8.6 ± 4.6
Acetic acid	9.9 ± 1.1	19.9 ± 2.3	26.1 ± 25.1
Allyl alcohol	1.7 ± 1.7	3.6 ± 3.6	6.1 ± 4.4
Amyl alcohol	7.6 ± 7.2	21.9 ± 20.8	20.5 ± 16.5
Butyric acid	15.1 ± 7.3	44.6 ± 21.5	59.4 ± 52.2
Diethylketone	5.7 ± 5.0	20.8 ± 11.4	11.0 ± 7.7
Ethyl acetate	8.6 ± 2.6	25.4 ± 4.8	12.7 ± 15.4
Ethyl alcohol	22.8 ± 10.0	44.7 ± 21.5	109.0 ± 31.5
Methyl alcohol	54.8 ± 29.3	74.4 ± 5.0	57.8 ± 6.3
Phenol	4.6 ± 1.9	9.5 ± 1.5	8.7 ± 5.3
Toluene	1.8 ± 1.7	7.4 ± 4.9	8.0
Inorganic Bioeffluent			
Carbon monoxide		$4.84 \times 10^3 \pm 1.2 \times 10^3$	
Ammonia		32.2 ± 5.0	
Hydrogen sulfide		2.73 ± 1.32	2.96 ± 0.68
Carbon dioxide		$642 \times 10^3 \pm 34 \times 10^3$	$930 \times 10^3 \pm 52 \times 10^3$
		$\dfrac{0.63 \text{ ft}^3 \text{ CO}_2^b}{\text{met} \cdot \text{hr} \cdot \text{person}}$	

[a]Means and standard deviations from Wang (1975).
[b]ASHRAE (1980). See Table 7.3 for relationship between mets and activity levels.

during and after 12-sec spray applications) represent "worst reasonable" conditions with two to four-fold variations with different products, ventilation conditions, and frequency and patterns of user application (Mokler et al., 1979b). The aerosol propellants were propane, isobutane, trichlorofluoromethane, and dichlorodifluoromethane, and all measurements were made on 1974 consumer products. It is perhaps important to note the *number median diameter* (NMD) (as opposed to the mass median diameter) for aerosol products is likely to be in the 0.04-0.06 μm diameter range (sizes measured for droplets from an insect spray can by Whitby et al., 1967). Particles of this size have high potential for deposition in the alveolar lung spaces.

Humans themselves constitute emission sources for various materials. Table 4.16 lists emission factors for a variety of organic and inorganic substances (Wang, 1975; ASHRAE, 1980). The quantitation of infectious bacteria and virus is less well developed. However, an emission rate of 93 infectious particles (quanta) per minute was estimated for an outbreak of measles in a Rochester school (Riley et al., 1978).

REFERENCES

Allan, G. G., Dutkiewicz, J., and Gilmartin, E. J. (1980). Long-term stability of urea-formaldehyde foam insulation. *Environ. Sci. Technol.* **14**:1235-1240.

Allen, R. J. (1977). *Relationship between indoor and outdoor concentrations of carbon monoxide and ozone for an urban hospital.* Ph.D. thesis, University of Illinois at the Medical Center, Chicago.

Allen, R. J., Wadden, R. A., and Ross, E. D. (1978). Characterization of potential indoor sources of ozone. *Am. Ind. Hyg. Assoc. J.* **39**:466-471.

Andersen, I., Lundqvist, G. R., and Molhave, L. (1975). Indoor air pollution due to chipboard used as a construction material. *Atmos. Environ.* **9**:1121-1127.

ASHRAE (1980). *Standards for ventilation required for minimum acceptable indoor air quality: ASHRAE62-73R.* American Society for Heating, Refrigerating and Air-Conditioning Engineers, New York.

Belles, F. E., Himmel, R. L., and DeWerth, D. W. (1975). Measurement and reduction of NO_x emission from natural gas-fired appliances. Paper 75.09.1, 68th Annual meeting of the Air Pollution Control Association, Boston, Massachusetts.

Berge, A. and Mellegaard, B. (1979). Formaldehyde emission from particle board— A new method of determination. *Forest Products J.* **29**(1):21-25.

Breysse, P. A. (1979). Formaldehyde exposure in mobilehomes and conventional homes. In: *Proceedings of the 43rd Annual Education Conference of the National Environmental Health Association*, June 23-28.

Breysse, P. A. (1981). The health cost of tight homes. *J. Am. Med. Assoc.* **245**: 267–268.

Butcher, S. S. and Buckley, D. I. (1977). A preliminary study of particulate emissions from small wood stoves. *J. Air Pollut. Control Assoc.* **27**: 346–348.

C&EN (1980). Mutagens found in photocopying toners. *Chem. & Eng. News*, 26, April 21.

C&EN (1982). Effects of foam insulation ban far reaching. *Chem. & Eng. News*, 34–37, March 29.

Chin-I-Lin, Anaclerio, R. N., Anthon, D. W., Fanning, L. Z., and Hollowell, C. D. (1979). *Indoor/outdoor measurements of formaldehyde and total aldehydes*, Lawrence Berkeley Laboratory Report, LBL-9397, July.

Cooper, J. A. (1980). Environmental impact of residential wood combustion emissions and its implications. *J. Air Pollut. Control Assoc.* **30**: 855–861.

Dally, K. A., Hanrahan, L. P., Woodbury, M. A., and Kanarek, M. S. (1981). Formaldehyde exposure in nonoccupational environments. *Arch. Environ. Health.* **36**: 277–284.

DeAngelis, D. G., Ruffin, D. S., and Reznik, R. B. (1979). *Source assessment: wood-fired residential combustion equipment field tests.* U.S. EPA Report No. EPA-600/2-79-019.

Dockery, D. W. and Spengler, J. D. (1981a). Personal exposure to respirable particles and sulfates. *J. Air Pollut. Control Assoc.* **31**: 153–159.

Dockery, D. W. and Spengler, J. D. (1981b). Indoor–outdoor relationships of respirable sulfates and particles. *Atmos. Environ.* **15**: 335–343.

Duncan, J. R., Morkin, K. M., and Schmierbach, M. P. (1980). Air quality potential from residential wood-burning stoves. Paper 80-7.2, 73rd Annual Meeting of the Air Pollution Control Association, Montreal, June 22–27.

FTC (1981). Federal Trade Commission. "Tar," nicotine, and carbon monoxide content of domestic cigarettes. *Fed. Regis.* **46**: 61828–61832, December 18.

Garry, V. E., Oatman, L., Pleus, R., and Gray, D. (1980). Formaldehyde in the home: Some environmental disease perspectives. *Minnesota Medicine* **63**: 107–111.

Geraci, C. L., Baron, P. A., Carter, J. W., and Smith, D. L. (1979). *Testing of hair dryers for asbestos emissions.* National Institute for Occupational Safety and Health, Cincinnati.

Hallenbeck, W. H. (1981). Consumer product safety: Risk assessment of exposure to asbestos emissions from hand-held air dryers. *Environ. Manage.* **5**: 23–32.

HEW (1979). *Smoking and Health–A Report of the Surgeon General.* Department of Health, Education and Welfare, Pub. No. (PHS) 79-50066.

Himmel, R. L. and Dewerth, D. W. (1974). *Evaluation of the pollutant emissions from gas-fired ranges.* Report No. 1492, American Gas Association Laboratories, Cleveland.

Hinds, W. C. (1978). Size characteristics of tobacco smoke. *Am. Ind. Hyg. Assoc. J.* **39**:48–54.

Hoegg, V. R. (1972). Cigarette smoke in closed spaces. *Environ. Health Perspect.* **2**:117–128, October.

Holcomb, M. L. and Scholz, R. C. (1981). *Evaluation of air cleaning and monitoring equipment used in recirculation systems.* NIOSH Pub. 81–113, National Institute for Safety and Health, Cincinnati, April.

Hollowell, C. D., Budnitz, R. J., Case, G. D., and Traynor, G. W. (1976). *Combustion-generated indoor air pollution, I. Field measurements 8/75–10/75.* Lawrence Berkeley Laboratories, Report LBL-4416, January.

Hollowell, C. D., Budnitz, R. J., and Traynor, G. W. (1977). Combustion-generated indoor air pollution. In: *Proceedings of the 4th International Clean Air Congress*, Tokyo, Japan, pp. 684–687.

Hollowell, C. D., Berk, J. V., and Traynor, G. W. (1979). Impact of reduced infiltration and ventilation on indoor air quality. *ASHRAE J.*, pp. 49–53, July.

Irving, K. F., Alexander, R. G., and Bavley, H. (1980). Asbestos exposure in Massachusetts public schools. *Am. Ind. Hyg. Assoc. J.* **41**:270–276.

Jenkins, R. A. and Gill, B. E. (1980). Determination of oxides of nitrogen (NO_x) in cigarette smoke by chemiluminescent analysis. *Anal. Chem.* **52**:925–928.

Kelley, T. F. (1965). Polonium-210 content of mainstream tobacco smoke. *Science* **149**:537–538.

Kusuda, T., Silberstein, S., and McNall, P. E., Jr. (1980). Modeling of radon and its daughter concentrations in ventilated spaces. *J. Air Pollut. Control Assoc.* **30**:1201–1207.

Long, K. R., Pierson, D. A., Brennan, S. T., Frank, C. W., and Hahne, R. A. (1979). Problems associated with the use of urea-formaldehyde foam for residential insulation. Part I: The effects of temperature and humidity on formaldehyde release from urea-formaldehyde foam insulation. University of Iowa. Published as ORNL/SUB-7559/I, Oak Ridge National Laboratory, U.S. Dept. of Energy, Oak Ridge, Tennessee, September.

Melia, R. J. W., Florey, C. duV., Darby, S. C., Palmes, E. D., and Goldstein, B. D., (1978). Differences in NO_2 levels in kitchens with gas or electric cookers. *Atmos. Environ.* **12**:1379–1381.

Mokler, B. V., Wong, B. A., and Snow, M. J. (1979a). Respirable particulates generated by pressurized consumer products. I. Experimental method and general characteristics. *Am. Ind. Hyg. Assoc. J.* **40**:330–338.

Mokler, B. V., Wong, B. A., and Snow, M. J. (1979b). Respirable particulates generated by pressurized consumer products. II. Influence of experimental conditions. *Am. Ind. Hyg. Assoc. J.* **40**:339–347.

NAS (1981a). *Formaldehyde and other Aldehydes*, National Academy of Sciences, Washington, D.C.

NAS (1981b). *Indoor Pollutants*, National Academy of Sciences, Washington, D.C.

NIOSH (1976). *Health hazard evaluation determination. Report No. 75-12-321. Cooper Union School of Art, New York, August 1976*, National Institute of Occupational Safety and Health, Cincinnati.

NIOSH (1977). *Health hazard evaluation/toxicity determination. Report No. 76-70-367, Sperry-Univac Corporation, Blue Bell, Pennsylvania.* National Institute for Occupational Safety and Health, March.

NIOSH (1979). *Health hazard evaluation. Report TA 79-047-825, Murray State University, Murray, Kentucky.* National Institute for Occupational Safety and Health, Cincinnati.

Nicholson, W. J., Swoszowski, E. J., Jr., Rohl, A. N., Todaro, J. D., and Adams, A. (1979). Asbestos contamination in United States schools from use of asbestos surfacing materials. *Ann. N.Y. Acad. Sci.* 330:587–596.

Palmes, E. D., Tomczyk, C., and DiMattio, J. (1977). Average NO_2 Concentration in dwellings with gas or electric stoves. *Atmos. Environ.* 11:869–872, 1977.

Palmes, E. D., Tomczyk, C., and March, A. (1979). Relationship of indoor NO_2 concentrations to use of unvented gas appliances. *J. Air Pollut. Control Assoc.* 29:392–393.

Riley, E. C., Murphy, G., and Riley, R. L. (1978). Airborne spread of measles in a suburban elementary school. *Am. J. Epidemiol.* 107:421–432.

Rossiter, W. J., Mathey, R. G., Burch, D. M., and Pierce, E. T. (1977). *Urea-formaldehyde based foam insulations: an assessment of their properties and performance.* Institute of Applied Technology, National Bureau of Standards, Washington, D.C.

Sardinas, A. V., Most, R. S., Giulietti, M. A., and Honchar, P. (1979). Health effects associated with urea-formaldehyde foam insulation in Connecticut. *J. Environ. Health* 41:270–272.

Sawyer, R. N. (1977). Asbestos exposure in a Yale building. *Environ. Res.* 13:146–169.

Sawyer, R. N. and Spooner, C. M. (1978). *Sprayed asbestos–containing materials: a guidance document.* U.S. EPA Report No. EPA-450/2-78-014, March.

Schmeltz, I., Hoffman, D., and Wynder, E. L. (1975). The influence of tobacco smoke on indoor atmospheres. I. An overview. *Preventive Medicine* 4:66–82.

Selway, M. D. (1979). *Characterization of photocopying machines as an indoor source of ozone.* M.S. Thesis, School of Public Health, University of Illinois, Chicago.

Selway, M. D., Allen, R. J., and Wadden, R. A. (1980). Ozone emissions from photocopying machines. *Am. Ind. Hyg. Assoc. J.* 41:455–459.

Spengler, J. D., Ferris, B. G., Dockery, D. W., and Speizer, F. E. (1979). Sulfur dioxide and nitrogen dioxide levels inside and outside homes and the implications on health effects research. *Environ. Sci. Technol.* 13:1276–1280.

Spengler, J. D., Dockery, D. W., Turner, W. A., Wolfson, J. M., and Ferris, B. G.,

Jr. (1981). Long-term measurements of respirable sulfates and particles inside and outside homes. *Atmos. Environ.* **15**: 23–30.

Sterling, T. D. and Sterling, E. (1979). Carbon monoxide in kitchens and homes with gas cookers. *J. Air Pollut. Control Assoc.* **29**: 238–241.

Stewart, R. D. and Hake, C. L. (1976). Paint-remover hazard. *J. Am. Med. Assoc.* **235**: 398–401.

Sutton, D. J., Nodolf, K. M., and Makino, K. K. (1976). Predicting ozone concentrations in residential structures. *ASHRAE J.*, pp. 21–26, September.

Thrasher, W. H. and Dewerth, D. W. (1975). *Evaluation of the pollutant emissions from gas-fired forced air furnaces.* Report No. 1503, American Gas Association Laboratories, Cleveland.

Thrasher, W. H. and Dewerth, D. W. (1977). *Evaluation of the pollutant emissions from gas-fired water heaters.* Report No. 1507, American Gas Association Laboratories, Cleveland.

UN (1977). *Sources and effects of ionizing radiation.* United Nations Scientific Committee on the Effects of Atomic Radiaton 1977 Report to the General Assembly, with Annexes, No. E77.1X.1, New York.

Wade, W. A., Cote, W. A., and Yocom, J. E. (1975). A study of indoor air quality. *J. Air Pollut. Control Assoc.* **25**: 933–939.

Wakeham, H. (1972). Recent trends in tobacco smoke research. In: *The Chemistry of Tobacco Smoke*, I. Schmeltz, Ed., Plenum Press, New York, pp. 1–20.

Wang, T. C. (1975). A study of bioeffluents in a college classroom. *ASHRAE Trans.* **81** (Part I): 32–44.

Weber, A., Fischer, T., and Grandjean, E. (1979). Passive smoking: irritating effects of the total smoke and the gas phase. *Int. Arch. Occup. Environ. Health* **43**: 183–193.

Whitby, K. T., Liu, B. Y. H., and McFarland, A. R. (1967). *A study of atomizer aerosol generators.* Progress Report No. AP 00480-2, University of Minnesota Particle Technology Laboratory, Minneapolis.

Yamanaka, S., Hirose, H., and Takada, S. (1979). Nitrogen oxides emissions from domestic kerosene-fired and gas-fired appliances. *Atmos. Environ.* **13**: 407–412.

Yocum, J. E. (1981). Combustion products. In: *Indoor Air Quality*, a report in preparation from the National Academy of Sciences. (Also see pp. 134–149 in reference NAS, 1981b.)

Zebransky, J. (Geomet Technologies, Inc.) (1981). *Proceedings: Conference on Wood Combustion Environmental Assessment (New Orleans, February 1981)* U.S. Environmental Protection Report EPA 600/59-81-029.

5

INDOOR MEASUREMENTS

Indoor measurements of air contaminants require consideration of a variety of factors. These include the selection of sampling equipment and an analytical technique with an adequate sensitivity, choice of a meaningful time scale for the measurement, the calibration of sampling and analytical methods, consideration of the need for area or personal sampling, evaluation of potential interferences with the analytical technique, and consideration of the effects of human activities on the concentration of the contaminant being measured. An understanding of how these relate to the process of air characterization is necessary for the proper measurement of air pollution in indoor environments.

SAMPLING AND ANALYSIS

Sampling methods fall into two general categories: dynamic methods requiring an air mover and passive methods based on diffusion or permeation to a collecting media. They also can be classified as (1) continuous or direct-reading methods, and (2) integrated or grab sampling. Continuous or direct-reading instruments combine sampling and analysis in one step. Integrated or grab sampling techniques require a separate analytical finish. The flow train for dynamic sampling typically consists of a sampling probe or inlet, flow controlling and/or flow measurement devices, gas or particle collector, and a pump or blower. These components are shown in Figure 5.1. Often several of these functions are combined in the same device. All standard reference methods are based on dynamic sampling.

A wide variety of sampling pumps are available (NIOSH, 1973). Personal sampling pumps are designed to operate at flow rates up to 4 liters/min for 8 hr and are powered by internal batteries. They include flow measurement and regulating devices that can be adjusted to meet the specific requirements of a sampling method, and are portable and relatively quiet. If a higher flow rate or longer sampling period is needed, a laboratory air pump (typically 28-140 liters/min) with flow control and measurement devices must be used.

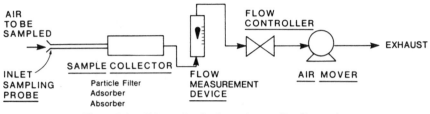

Figure 5.1.   Schematic of a dynamic sampling flow train.

One popular flow controller is the critical flow orifice. The device is designed so that sonic velocities are achieved in the orifice. Consequently, if the downstream to upstream pressure ratio is less than a critical value (~0.5 for a circular knife-edged design) and the upstream pressure is close to 1 atm, the flow rate through the orifice will remain constant. Another system used to regulate and measure flow consists of a rotameter (a device for measuring air flow rate) and a pressure regulator, needle valve, or by-pass valve. Special care is required to calibrate any of these systems and will be discussed under *Calibration*.

Particles

A particle collection device generally consists of a filter, flow controller, and a pump. A variety of filter media are available for particle sampling. Considerations in selecting a filter include collection efficiency (John and Reischl, 1978; Liu et al., 1978; Liu and Lee, 1976), pressure drop, background contamination of the substance of interest, and compatibility of the filter media with the analytical method. For example, cellulose filters are often recommended for the sampling of metals. These filters are cheap, reasonably efficient, and have very low background levels of most metals. Unfortunately, cellulose filters are very hydroscopic, making gravimetric determinations of total particulate matter very difficult (Neustadter et al., 1975). Glass or quartz fiber filters are not hydroscopic but can have significantly higher background metal concentrations. Teflon or polyvinyl chloride membrane filters are popular because of their low background and nonhydroscopic qualities. These filters are more expensive and can have very high pressure drops requiring more powerful pumps (John and Reischl, 1978). Unfortunately, there is no one filter best suited for all applications. In general, cellulose, fiber, and membrane filters have good collection efficiencies and are appropriate to use for collecting indoor particle samples. Under special conditions where submicron particles are present and represent a potential hazard, a high-efficiency filter should be used. A small pore membrane or fiber filter would be a good choice for this application. If pressure drop characteristics of filter media are available, filters with the higher pressure drops will generally have higher efficiencies for submicron particles.

When sampling for total particulate matter, filter surfaces should be left open to the environment being sampled. An inlet tube before a particle filter can cause significant errors from particle loss to the tube's inner surface from diffusion and impaction (El-Shobokshy and Ismail, 1980). If an inlet tube is necessary, it is desirable to keep the tube as straight and short as possible.

Size is an extremely important particle characteristic (e.g., see Willeke and Whitby, 1975). The size of a particle suggests something about its source, potential toxic affect, and residence time in the atmosphere. Samplers are available that collect size-segregated particle samples. The cyclone preseparator is a size-selective device designed to operate with a personal sampling pump and standard filter cartridge (Aerosol Technology Committee, 1970). The cyclone sampler essentially removes the nonrespirable fraction of the total particulate matter entering the device before the air stream reaches the filter. A respirable particle is defined as one which is capable of reaching the noncilliated portion of the lung (Miller et al., 1979). This corresponds to particles with an aerodynamic diameter $<2.5~\mu m$. The particle filter placed after a cyclone preseparator collects essentially all particles with aerodynamic diameters $<2.5~\mu m$.

Another type of size-selective sampler is the cascade impactor (e.g., Allen and Wadden, 1977; Willeke, 1975; Marple and Willeke, 1976). These devices separate particles into four or more size ranges and operate at flow rates from <28 liters/min (1.0 cfm) to up to $1.1~m^3$/min (40 cfm). Special care is required to make sure the flow rate is as specified by the manufacturer and is constant throughout the sampling period. Fluctuations can affect the size-selective characteristics of the impactor. Electronic flow controllers are available to control this parameter. Placing a high-volume sampler (such as $1~m^3$/min cascade impactor) in a small room can significantly change the characteristics of pollution in the room by blowing dust and changing air infiltration rates (as well as considerably raising the noise level). It is a good idea to exhaust a high-volume blower outside of the room being evaluated.

A number of direct-reading particle samplers are currently available. The Electrical Aerosol Analyzer is an electrostatic sampler designed to measure the particle number distribution between 0.0032 and 1.0 μm in ten size intervals (Liu and Pui, 1975). Optical particle counters (e.g., Royco 245) measure number concentration in various ranges from 0.3 to 10 μm diameter (ACGIH, 1978). The integrating nephelometer measures light scattering due to particles in the air (Waggoner and Charlson, 1977; Charlson et al., 1969). Output of the nephelometer is highly correlated with respirable particle concentration (e.g., Scheff and Wadden, 1979) and can operate for long periods of time unattended. The condensation nuclei counter measures condensation nuclei concentration (Lui and Pui, 1974). The instrument is most sensitive to particles in the 0.01-0.1 μm size range. Piezoelectric devices provide a direct reading of suspended mass in the 0.01-3.0 μm range. This design is subject to nonlinear effects from long sampling periods and is difficult to operate in the field because of crystal cleaning requirements and humidity and temperature effects (Lundgren et al., 1976). However, when properly calibrated, such devices are reported to be sensitive to less than 10 $\mu g/m^3$ per minute (ACGIH, 1978). In a controlled experiment on the characteristics of tobacco smoke, a piezobalance underestimated the mass concentration by 15% compared to low-volume filter sampling techniques (Repace and Lowrey, 1980). Table 5.1 lists characteristics of a variety of direct-reading particle monitors (ACGIH, 1978).

Gases

Dynamic gas collection devices either absorb the sample in a liquid medium in an impinger, or adsorb the sample on a solid surface. Most common standard methods are based on liquid-filled impingers. However, solid adsorbing media are becoming more popular. Factors of concern in gas sampling are collection efficiency (absorption) and breakthrough (adsorption). If the collection efficiency or breakthrough of a particular method is not known, a simple way to evaluate this factor is to place a second absorber or adsorber downstream of the first. The collection efficiency or breakthrough can then be estimated by dividing the mass collected by the first absorber by the mass collected by both absorbers. Because the second absorber is rarely 100% efficient, this calculation will only be an estimate. Collection efficiency can frequently be increased by (1) decreasing flow rate; (2) using a fritted bubbler in an impinger; (3) increasing the length of an adsorbing bed; and (4) using two or three collectors in series.

Increasing sampling flow rate can increase the sensitivity of the measurement by providing more sample for the analytical finish. Caution must be exercised when using this approach. Many analytical methods have recommended sampling flow rate and duration ranges. These limits are designed to ensure efficient collection and to avoid loss or decay of sample during long sampling periods. Such recommendations must be followed or at least evaluated when selecting an appropriate flow rate and duration. Analytical sensitivity cannot be increased too much by modifying sampling parameters without paying the price of decreased efficiency or sample loss. This is especially important for gas collection devices.

A variety of direct-reading gas monitors are currently available. A number of these are sensitive enough for nonindustrial indoor applications. Ultraviolet detectors for ozone are commercially available with a detection limit of 2 ppb (EPA, 1978). Monitors for NO and NO_x that operate on the principle of a chemiluminescent reaction of NO with ozone (e.g., Monitor Labs, 1975) can respond to concentrations as low as 5 ppb. Nondispersive infrared analyzers for CO are available with detection limits as low as 0.1 ppm (EPA, 1971). Portable electrochemical detectors for CO are available with detection limits

Table 5.1 Characteristics of Direct-Reading Particle Monitors

Instrument	Analytical Method	Type of Measurement	Effective Particle Size Range	Portable
Aerosol Photometers	Light scatter at a fixed angle	Indirect measure of aerosol mass or dust	<1.0–10.0 μm	Yes
Integrating nephelometer	Light scatter integrated over a wide angle ($8°$–$170°$)	Visual range: indirect measure of respirable particles	0.1–1.0 μm	No
Light scattering optical counters	Light scatter at a fixed angle	Particle number distribution	0.3–10.0 μm in up to 10 size ranges	Yes
Active scattering aerosol spectrometer	Light scatter of output of a He–Ne laser at a fixed angle	Particle number distribution	0.08–4.5 μm in up to 48 size ranges	No
Condensation nuclei counter	Condensation of nuclei by adiabatic expansion and optical detection	Condensation nuclei concentration	0.0025–1.0 μm	Yes
Electrical aerosol analyzer	Electrostatic particle mobility analyzer	Particle number distribution	0.0032–1.0 μm in 10 ranges	No
Respirable dust mass monitor	β-attenuation	Cyclone followed by an indirect measure of mass	<10 μm	Yes
Aerosol mass monitor	Piezoelectric crystal	Particle mass concentration	0.01–10.0 μm	Yes

83

of 1.0 ppm (Bay et al., 1972). Many of the direct-reading gas-measuring instruments are designed to operate unattended for long periods of time. Table 5.2 lists the various types of direct-reading gas monitors. Of note is that commercially available, portable systems for the measurement of ambient levels of CO_2 (300–600 ppm) are not currently well developed.

Another area of concern when designing a gas sampling system is the proper selection of materials. In general, glass or Teflon tubing should be used upstream of the collector. This is especially important for reactive gases like ozone. A short piece of polyethylene tubing upstream of an ozone collector or detector can cause a significantly low measurement.

The use of passive samplers for the collection of gaseous pollutants is an area of current research. Because these samplers have no pumps, they are very compact and lightweight. This can be an important factor when trying to measure many pollutants simultaneously. Most such samplers are based either on diffusion-controlled collection with an adsorbing or absorbing substance, or on transfer through a permeable membrane to a collecting medium. Figure 5.2 is a schematic of the two types of processes.

Table 5.2   Summary of Methods for Direct-Reading Gas Monitors

Gas	Analytical Methods	Sensitivity (ppm)	Portable
Carbon monoxide	Electrochemical oxidation	1.0	Yes
	Gas chromatography	0.1	No
	Infrared photometry	0.1	No
Ozone	Coulometry	0.003	Yes
	Chemiluminescence	0.001	No
	Ultraviolet photometry	0.002	No
Sulfur dioxide	Coulometry	0.01	No
	Colorimetry	0.005	No
	Chemiluminescence	0.01	No
	Gas chromatography	0.01	No
Nitric oxide	Oxidation to NO_2/ colorimetry	0.002	No
	Chemiluminescence	0.005	No
Nitrogen dioxide	Colorimetry	0.002	No
	Conversion to NO/ chemiluminescence	0.005	No
Hydrocarbons	Gas chromatography	0.01	No
Carbon dioxide	Infrared photometry	0.5%	No

Diffusion-controlled samplers are based on Fick's Law:

$$J = -D \frac{dC_g}{dx} \tag{5.1}$$

where J is the flux of pollutant, mass/area · time, D is the gas diffusivity, area/time, and dC_g/dx is the gas concentration gradient, mass/volume · distance. The concentration gradient is caused by absorption or adsorption of the pollutant at the collecting surface. For constant diffusivity, a straight-line concentration gradient, and zero concentration at the collecting surface, the average concentration of the pollutant over a collection period t will be

$$C_g = \frac{m\Delta x}{tDA} \tag{5.2}$$

where m is the mass of pollutant collected, Δx is the length of the diffusion path, and A is the contact area of the collecting medium. This approach has been suggested for measurement of NO_2 (Palmes et al., 1976; Tompkins and Goldsmith, 1977), NO_x (Palmes and Tomczyk, 1979), SO_2 (Palmes and

DIFFUSION-CONTROLLED PROCESS

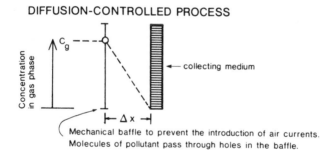

Mechanical baffle to prevent the introduction of air currents. Molecules of pollutant pass through holes in the baffle.

PERMEATION-CONTROLLED PROCESS

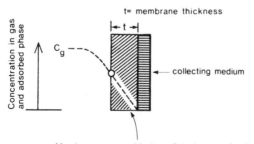

Membrane permeable to pollutant gas molecules. Pollutant dissolves in the membrane.

Figure 5.2.   Passive monitoring processes.

Gunnison, 1973; Tompkins and Goldsmith, 1977), CO and HCN (Obermayer et al., 1980), and a variety of organic vapors (Bamberger et al. 1978; Orofino and Usmani, 1980; Lautenberger et al., 1980; Hickey and Bishop, 1981; Hill and Fraser, 1980; Evans et al., 1977; Gillespie and Daniel, 1979). Difficulties with this design include the effect of convection currents on the concentration gradient, dependence of D on concentration, dosimeter geometry, effects of temperature, and incomplete collection in the absorbing or adsorbing medium.

For permeation-controlled collectors, the pollutant is dissolved in a polymeric membrane, transported across the membrane because of the concentration gradient between the gas-contact interface and the dissolution surface, and absorbed or adsorbed into or onto an appropriate chemical or surface. The time-weighted average concentration is then

$$C_g = \frac{mk}{t} \qquad (5.3)$$

where k is the permeation constant for the particular pollutant, polymer, and collector geometry. Devices have been developed which measure CO (Bell et al., 1975), hydrogen cyanide (West, 1980; Amass, 1979), vinyl chloride (Nelms et al., 1977; West and Reiszner, 1978), hydrogen sulfide (Hardy et al., 1981; Amass, 1979; West, 1980), chlorine (Hardy et al., 1979), sulfur dioxide (Reiszner and West, 1973; Amass, 1979), ammonia (West, 1980; Amass, 1979), and a few organics (West and Reiszner, 1979). Film thickness, other airborne substances, temperature, and relative humidity may all affect membrane response, and overall efficiency will be further affected by the capacity, efficiency, and temperature dependence of the collection medium.

Passive sampling systems require some type of analytical finish. In the simplest case, this can be reading the color change of an indicating substance (Hill and Fraser, 1980). Most passive systems for inorganics require a wet-chemical spectrophotometric analytical finish (e.g., Palmes et al., 1976). Passive sampling for organics generally involves chemical desorption followed by gas chromatography (e.g., Hickey and Bishop, 1981).

Commercially available passive sampling systems represent a new development in the fields of industrial hygiene and environmental measurement. Very little performance data on these systems are available. While many passive monitoring systems have been calibrated with laboratory-generated gas mixtures, their reliability, accuracy, and response to interferences under field conditions where they may be exposed to harsh environments and complex mixtures of gases are not well defined. It is therefore necessary to calibrate and field test a passive sampling system before using it to collect environmental data. The field test will ordinarily include a comparison of passive sampler measurements with measurements by a reference method. The reference method should be a standard method with known performance and interference characteristics. Samples should be collected in reference–passive pairs and the distribution of the differ-

Table 5.3 Some Commercially Available Passive Sampling Systems[a]

Sampler/Manufacturer	Components Measured	Mass Collection Sensitivity	References
PRO-TEK samplers	Organic vapors	[b]	Lautenberger et al. (1980)
E.I. DuPont	Benzene	0.2 ppm · 8 hr	EPA (1981)
Applied Technology	NO_2	1 ppm · hr	E.I. DuPont (1981)
Division	SO_2	1 ppm · hr	
	NH_3	5 ppm · hr	
	HCHO	2 ppm · hr	
Abcor GASBADGE	Organic vapors	[b]	Gillespie and Daniel (1979)
Abcor Development	Benzene	0.2 ppm · hr	Bamberger et al. (1978)
Corporation			Tompkins and Goldsmith (1977)
3M Brand #3300	Organic vapors	[b]	Hickey and Bishop (1981)
3M Model 3750	HCHO	0.77 ppm · hr	Wallace and Ott (1982)
Occupational Health &			
Safety Products Division			
3M Company			
Mini Monitor	SO_2	0.01 ppm · 8 hr	West and Reiszner (1979)
Reiszner Environmental	Cl_2	0.013 ppm · 8 hr	
& Analytical Labs,	Alkyl lead	5 $\mu g/m^3$ · 8 hr	Nelms et al. (1977)
Inc. (REAL)	Vinyl chloride	0.02 ppm · 8 hr	West and Reiszner (1978)
	Benzene	0.02 ppm · 8 hr	
	NH_3	0.4 ppm · 8 hr	West (1980)
	H_2S	0.01 ppm · 8 hr	
	HCN	0.01 ppm · 8 hr	
Monitox System	HCN	TLV: 10 ppm · 8 hr	MDA (1981)
MDA Scientific Inc.	NO_2	TLV: 5 ppm · 8 hr	EPA (1981)
	Phosgene	TLV: 0.1 ppm · 8 hr	
	H_2S	TLV: 10 ppm · 8 hr	
	CO	TLV: 50 ppm · 8 hr	
Leak-Tec Personnel	NH_3	>25 ppm · 8 hr	American Gas & Chemical (1981)
Protection Indicators	CO	>50 ppm · 8 hr	EPA (1981)
American Gas &	Cl_2	>2 ppm · 8 hr	
Chemical Co.	Hydrazine	>5 ppm · 8 hr	
	H_2S	>5 ppm · 8 hr	
	NO_2	>1 ppm · 8 hr	
	O_3	>1 ppm · 8 hr	
SKC Monitoring Badge	Phosgene	0.1 ppm · hr	SKC (1981)
SKC Inc.			EPA (1981)
Model 530 NO_2/NO_x	NO_2	1 ppm · hr	Palmes et al. (1976)
Field Kit			
MDA Scientific, Inc.	NO_x	1 ppm · hr	Palmes and Tomczyk (1979)

[a]Mention of specific products or manufacturers does not constitute endorsement of these items.
[b]Sensitivity varies for each compound.

ences tested statistically. Table 5.3 lists some commercially available passive sampling systems.

SELECTION OF SAMPLING AND ANALYTICAL METHODS

Most analytical methods require samples to be collected in a specific way. Two basic types of measurement techniques are available. Those that combine sampling and analysis continuously are the direct-reading methods. The electrochemical oxidation cell for the measurement of CO is a common example (Bay et al., 1972). The other basic type of measurement technique involves the sampling of air to remove the contaminant of interest followed by the chemical analysis of the collected sample in the laboratory. Bubbler and filter methods fall into this category.

Many considerations go into the selection of a measurement method. These include the availability of a standard method, availability of proper instrumentation, cost of analysis, sensitivity and time scale required for the measurement, and the effects of sampling duration and interferences on the analytical method. When available, standard methods should always be used. There are a number of good references for standard methods for the analysis of air. These include EPA criteria documents (e.g., EPA, 1978b), *Methods of Air Sampling and Analysis* (APHA, 1977), the National Institute for Occupational Safety and Health's *Manual of Analytical Methods*, Vol. I–III (NIOSH, 1977), *Measuring, Monitoring and Surveillance of Air Pollution*, Volume 3 (Stern, 1976), and *Standard Methods for the Examination of Water and Wastewater* (APHA, 1975).

An estimate of the minimum concentration that one expects to measure is very useful in evaluating the appropriateness of an analytical method. The minimum concentration is directly related to the mass that can be collected for a given time period and sampling rate. This minimum mass, in turn, determines the sensitivity required of an analytical method. An examination of the literature can often provide enough data, collected under conditions similar to the application at hand, to allow a concentration estimate. For example, if one is called on to measure the impact of cigarette smoking on indoor levels of nicotine and methylchloride, data on the impact of cigarette smoking on indoor air quality can be found in the report of the Surgeon General on smoking and health (HEW, 1979). This review suggests that levels of nicotine >1 $\mu g/m^3$ are likely to be found in any tobacco smoke polluted indoor environment, even under well-ventilated conditions. If a personal sampling pump is to be used with an open-face filter cassette at a flow rate of 2 liters/min for 4 hr, the minimum mass of nicotine collected on each filter will be

$$\text{Nicotine collected} \geqslant \left(\frac{1\ \mu g}{m^3}\right) \left(\frac{m^3}{10^3\ \text{liters}}\right) (4\ \text{hr}) \left(\frac{60\ \text{min}}{\text{hr}}\right) \left(\frac{2\ \text{liters}}{\text{min}}\right)$$

$$\geqslant 0.48\ \mu g \tag{5.4}$$

Any analytical technique for the determination of nicotine that requires less than 0.48 μg could be used for this application.

If indoor concentration data are not available for the above calculation, as is the case for methylchloride (CH$_3$Cl), emission data can frequently be used to estimate concentration using a simple ratio model and the concentration of a well-characterized pollutant. An estimate of the expected concentration of CH$_3$Cl can be made from emission data for CH$_3$Cl and CO, and data on the indoor concentration of CO from cigarette smoke. The decay and dispersion characteristics of the two compounds are assumed not to be radically different. Table 4.5 gives average emissions of 37.5 mg CO and 1.3 mg CH$_3$Cl per cigarette in sidestream smoke (the fraction of cigarette smoke directly emitted from the tip of a burning cigarette). The Surgeon General's report indicates that CO concentrations where smokers are present are always >1 mg/m^3. Therefore, an estimate of the minimum expected concentration of CH$_3$Cl is

$$\text{Methylchloride concentration} \geqslant \left(\frac{1 \text{ mg CO}}{\text{m}^3}\right)\left(\frac{\text{cigarette}}{37.5 \text{ mg CO}}\right)\left(\frac{1.3 \text{ mg CH}_3\text{Cl}}{\text{cigarette}}\right)$$

$$\geqslant 0.035 \text{ mg/m}^3 \tag{5.5}$$

This air concentration can be directly related to the analytical sensitivity of a direct-reading instrument for CH$_3$Cl, or used as in Equation (5.4) to estimate the minimum collectable mass. Other methods for estimating the indoor concentration of pollutants are given in the section on *Indoor Air Quality Prediction.*

Many methods for the analysis of gaseous pollutants involve the collection of the pollutant in a liquid-filled impinger. The detection limit of these methods are generally expressed as μg pollutant per ml absorbing solution. If this type of method is selected to measure CH$_3$Cl for the example described above, with sampling at 0.5 liter/min for 4 hr using 10 ml of absorbing solution and 90% collection efficiency, an estimate of the minimum mass of methylchloride per ml absorbing solution will be

$$\text{Methylchloride collected} \geqslant \left(\frac{35 \ \mu\text{g CH}_3\text{Cl}}{\text{m}^3}\right)\left(\frac{\text{m}^3}{10^3 \text{ liters}}\right)(4 \text{ hr})(0.9)$$

$$\cdot \left(\frac{60 \text{ min}}{\text{hr}}\right)\left(\frac{1}{10 \text{ ml absorbing solution}}\right)\left(\frac{0.5 \text{ liter}}{\text{min}}\right)$$

$$\geqslant 0.38 \ \frac{\mu\text{g CH}_3\text{Cl}}{\text{ml absorbing solution}} \tag{5.6}$$

Therefore, any analytical method with a detection limit less than 0.38 μg CH$_3$Cl per ml of absorbing solution would be adequate to measure the lowest concentration expected in an indoor environment contaminated by cigarette smoke.

As an example of passive dosimeter application, Table 5.4 shows estimates of

Table 5.4   Indoor Concentration and Estimated Performance of Passive Samplers for Seven Tobacco Smoke Constituents

Constituent	Sidestream Emissions[a] (μg/cigarette)	Measured or Predicted Concentration[b] ($\mu g/m^3$)	Diffusivity (25°C) (cm^2/s)	Mass Collected by Passive Sampler[c] (μg)
HCN	110	67	0.15	0.029-0.29
Acrolein	180	110	0.087	0.027-0.27
Acetaldehyde	1570	960	0.087	0.24-2.4
Formaldehyde	1300	790	0.087	0.20-2.0
NO	2300	1,400	0.20	0.81-8.1
NO_2	625	380	0.14	0.15-1.5
CO	37,500	22,900 (20 ppm)	0.20	13.2-132

[a]From Table 4.5.
[b]Equation (5.5).
[c]Equation (5.2); $0.1 \leqslant A/\Delta x \leqslant 1.0$; t = 8 hr.

indoor concentration, diffusivity, and mass collected by diffusion-controlled collectors for seven gaseous components of tobacco smoke. Diffusivities were calculated from molecular theory (e.g., Reid and Sherwood, 1966) at 25°C. Area to length ratios for diffusion collectors of 0.1-1.0 are used. Average side stream emissions are from Table 4.5 and indoor concentrations of HCN, acrolein, acetaldehyde, NO, and NO_x are based on a ratio model with CO [e.g., Equation (5.5)]. Mass collected by the passive samplers is based on 8-hr sampling.

When evaluating the appropriateness of an analytical method, one is often faced with the problem that both indoor concentration data and emission data for the substance of interest are not available. In this case, a different approach based on the potential health effect level can be used for determining the required sensitivity. The analytical method should be capable of measuring 10% of the minimum health effect level or environmental standard.

One undesirable side-effect of electrostatic precipitators (devices used to remove particulate matter from air) can be the production of ozone (see Table 4.12). The EPA primary health standard for ozone is 0.12 ppm as a maximum hourly average. To evaluate a possible ozone problem resulting from the use of an electrostatic precipitator, the method selected should have a minimum detectable limit of $\leqslant$10% of 0.12 or 0.012 ppm ozone.

Even after going through the exercise of selecting an analytical method with an acceptable sensitivity, it is still possible to encounter concentrations below the detection limit. It is important to understand that such a measurement is not invalid or incorrect, but only that the concentration during the sampling period was less than the detection limit of the analytical method. If, for example, ten measurements of formaldehyde are made in a home and three are below the detection limit, an average of the seven measurements above the detection limit is not a valid average concentration. The three measurements below the detection limit should also be considered. One useful way to handle this type of problem is to calculate two averages, a maximum average and a minimum average. The maximum average assumes the three measurements below the detection limit are equal to the detection limit, and the minimum average assumes the three measurements are equal to zero. The actual average concentration of formaldehyde will lie between these maximum and minimum values.

CALIBRATION

All systems for sampling and analysis of air pollution must be calibrated before they can be used. This generally involves the calibration of the sample collection system as well as the chemical analysis procedures. The amount of sample, whether collected by a filter, an adsorber or absorber, or indicated by a direct-reading instrument, depends on the volume of air sampled. Many instruments are available to calibrate the air flow characteristics of a sampling system. A soap bubble meter is useful for measuring flow rates less than 1 liter/min (NIOSH, 1973). The meter can be made from an inverted laboratory volumetric buret. When the interior surfaces are wetted with a soap solution, a soap bubble formed on the bottom of the tube, and suction applied to the top, the bubble will be drawn up to the top of the buret. The distance the bubble moves will be proportional to the total volume of air pumped. Therefore, by measuring the time it takes the bubble to move a known distance, the volumetric flow rate can be calculated. This method is very useful for accurately measuring very low flow rates (<100 ml/min).

The wet-test meter is useful for calibrating flow rates in the 0.5–20 liters/min range (ACGIH, 1978). The meter works on the principle of displacing a known volume of water to rotate an indicator dial. For accurate results, the meter must be level and filled with the correct amount of water. The pressure difference between the flow system being calibrated and the atmosphere cannot exceed 6 inches of H_2O. The volume of air measured by a wet-test meter is always on a wet basis. This can be corrected to standard conditions as follows:

$$V_{STP} = \frac{T_s}{T_m} \left[\frac{P_m - (P_{H_2O} + \Delta P)}{P_s} \right] V_m \qquad (5.7)$$

where V_{STP} = volume at 25°C, 1 atm
$\quad\quad\ T_s = 298°K$

T_m = temperature measured, °K

P_m = atmospheric pressure, atm

P_{H_2O} = partial pressure of H_2O at 100% RH, atm

ΔP = pressure drop across the wet test meter, atm

P_s = 1 atm

V_m = volume measured by wet test meter.

Accuracy of the wet-test meter is within 1% (NIOSH, 1973).

For higher flows a dry drum gas meter can be used. The dry drum meter consists of two chambers, separated by a diaphragm, interconnected by mechanical valves and a cycle-counting device. These devices are useful for flows up to 8.5 m³/hr (5 ft³/min) (ACGIH, 1978). The mechanics of the meter can cause a significant pressure drop that must be corrected for when using it to calibrate an air sampler. Equation (5.7) can be used to correct the volume measured to STP. In this case, however, P_{H_2O} will be the actual partial pressure of water at the existing relative humidity and ΔP will be the pressure drop across the dry drum meter. Orifice type calibrators should be used for calibrating high-volume devices which operate at 34–100 m³/hr (20–60 ft³/min). These devices operate on the principle that the flow through the orifice is proportional to the pressure drop across it (EPA, 1971).

Many commercial air samplers use a rotameter as a direct-reading indicator of air flow rate through the system. A rotameter is a variable area meter with a float that moves up and down in a vertical tapered tube. The tube cross-sectional area is larger at the top than the bottom. As flow is increased through the rotameter, the float will move to a higher position in the tube. Rotameters have a range of about a factor of 10 and can measure flow ranges from a few cm³/min to over 30 m³/min (NIOSH, 1973).

All rotameters come with some type of scale. The scale can be a unitless index or have flow rate units printed on it. In either case, all rotameters must be calibrated before they are to be used to measure air flow. Rotameters are very sensitive to pressure changes in a sampling system. Consequently, it is necessary to calibrate them in place with the sampling system exactly as it will be used during operation. If, for example, a rotameter is to be used downstream of an impinger, the impinger must be filled with absorbing solution and in the system during calibration.

Flow through a critical flow orifice, on the other hand, is only a function of upstream conditions as long as the critical pressure drop across the device is maintained. Once calibrated, these devices are relatively insensitive to pressure changes and can maintain constant flow for long periods of time.

Systems that generate known concentrations of air contaminants are useful for calibrating direct-reading instruments and for testing or verifying integrated methods. Measurements of a range of known concentrations by a direct-reading pollutant monitor, for example, can be used to develop a calibration curve for instrument response over a concentration range of interest. Calibrated instruments are strictly applicable only to air concentrations within the calibration range.

Two accurate ways of generating known concentrations of gaseous pollutants are with permeation tubes and calibrated gas mixtures (ACGIH, 1978). A permeation tube is a Teflon cylinder filled with liquified pollutant. As long as the tube is held at a constant temperature and some of the pollutant remains in the liquid phase, gas molecules will permeate out of the tube at a constant rate. A low flow of dry nitrogen (or other inert carrier gas) is passed over the tube in a constant temperature oven or water bath. Adjusting the relative flow rates of clean dilution gas and carrier gas into a mixing chamber will vary the calibration gas concentration. Permeation tubes are relatively inexpensive, last several months (much longer if stored in a refrigerator between calibrations), and the mass flow rate discharged can be calculated by measuring the change of weight of the tube over time. Gases for which permeation tubes are available include SO_2, NO_2, acrolein, formaldehyde, HF, H_2S, and NH_3 (e.g., Metronics Associates, Palo Alto, Calif.).

Calibrated cylinder gas mixtures are an accurate way of generating standard concentrations for many substances for which permeation tubes are not available. Gas mixtures such as methane in air or CO in air can be stored for long periods without changes in concentration. In most cases, flow dilution systems are needed to reduce the concentration of the cylinder gas to a range of interest. There are a number of important factors to consider when designing a gas dilution system. Very high dilution ratios require a very low-flow pollutant gas stream to be mixed into a high-flow clean air stream. Because of the difficulty of calibrating a low-flow system, dilutions of greater than 500:1 should be avoided. Many flow measuring and regulating devices are sensitive to changes in pressure. It is important to keep the pressure drop in the system low and operate the system close to atmospheric pressure. All flow measuring and regulating devices in a dilution system must be calibrated in place under normal operating conditions. This will help control errors from pressure changes. It is also desirable to use only glass or Teflon parts in a calibration flow train. Polyethylene and metal tubing may require several hours of conditioning before they stop adsorbing or desorbing the test substance. Finally, clean air should be used for the dilution air stream. Background concentrations of some pollutants can cause a significant error when generating low concentrations. A drying bed (silica gel), ozonizer (ultraviolet lamp), charcoal trap, and particle filter in series form an excellent clean-up system. The silica gel, charcoal, and particle filter should be replaced as needed.

The generation of standard aerosol mixtures is somewhat more difficult than that of gas mixtures. Aerosol generators are available that can generate a polydisperse particle size distribution (e.g., compressed air nebulizer) or a mono-disperse distribution (e.g., vibrating orifice, spinning disk, and nebulization of monodisperse suspensions of polystyrene spheres) (ACGIH, 1978; Liu and Pui, 1975; NIOSH, 1973; Kim et al., 1981; Raabe, 1976). These generators are useful for evaluating the particle sizing characteristics of aerosol analyzers. An ultra-sonic nebulizer can be used to generate log-normal distributions of sulfate aerosol with stable ($\pm 5\%$ over 6 hr) size characteristics and mass concentrations (Kim et al., 1981). The mass median aerodynamic diameter of the aerosol range from 0.3 to 2.29 μm (for sulfate salt solutions of 0.01–10.0% by weight, respectively) with geometric standard deviations of 1.4–1.6.

For further discussions on the design of calibration systems see *The Industrial Environment–Its Evaluation and Control* (NIOSH, 1973), *Air Pollution*, Volume III (Stern, 1976), *Methods of Air Sampling and Analysis* (APHA, 1977), and *Air Sampling Instruments for Evaluation of Atmospheric Contaminants* (ACGIH, 1978).

TIME SCALE

The time scale of a measurement should be related to the potential health effect of the pollutant of concern, as well as the conditions of exposure. Integrated sampling methods measure the average concentration of a pollutant during the sampling period. These methods are useful for characterizing exposures to pollutants with toxic effects resulting from long-term exposure. Direct-reading methods measure the instantaneous concentration of a pollutant. These methods are most appropriate for characterizing exposures to pollutants with short-term toxic effects. By continuously recording and integrating the output of direct-reading instruments, such instruments can also be used for integrated sampling.

An evaluation of occupational and environmental standards can help determine which type of sampling is most appropriate. If the pollutant of concern has an occupationally related short-term exposure limit (STEL) or an hourly environmental standard, it has significant short-term toxic effects and peak levels or maximum hourly averages are of concern. If available, a continuous direct-reading instrument should be used. Unfortunately, direct-reading instruments are usually considerably more expensive than integrated methods and are only available for a limited number of substances. When a direct-reading method is not available for measuring peak concentration, a series of short collection-period samples using an integrated sampling method with a low detection limit can be used. An estimate of the amount of mass expected to be collected during short sampling times is needed to be sure of success in measuring the pollutant of concern.

Substances which do not have an occupational STEL or have environmental standards with 24-hr or annual average exposure limits can be adequately characterized by integrated sampling methods. The important factor here is to sample during the entire exposure period. For example, if exposure to a pollutant of concern is suspected in an office building, sampling for a full 8-hr work-shift will provide a better measure of the potential toxic effect than a grab sample or short-term sample.

There are no fixed rules when determining the time scale of a measurement. It is, therefore, necessary to evaluate the toxic effects of the pollutant of concern and the conditions of exposure to determine if a method selected for a specific application is appropriate. The calculations described in the section on *Selection of Sampling and Analytical Methods* are useful for this evaluation.

INTERFERENCES

When measuring the concentration of a pollutant, interference from other chemical species present can often result in significant errors. Most standard methods include a list of major interferences. The concentration of chemical species known to interfere should be evaluated (measured or estimated) when selecting and using an analytical method. The error caused by the interference can frequently be estimated and, depending on its magnitude, corrected for or neglected.

The neutral buffered potassium iodide method has been commonly used for measuring ozone in the 0.01-10 ppm range (APHA, 1977). The method lists the following interferences: sulfur dioxide produces an equimolar negative interference; nitrogen dioxide is equivalent to a positive 10% of an equimolar concentration of ozone; and peroxyacetyl nitrate (PAN) is equivalent to a positive 50% of an equimolar concentration of ozone. Given this type of information, an evaluation of how to apply the neutral buffered KI method to a variety of situations can be made. For example, for outdoor atmospheric measurements, the effects of SO_2, NO_2, and PAN cannot be neglected. If the measured concentration of ozone using the neutral buffered method is 0.15 ppm and, during the sampling period, the concentrations of SO_2, NO_2, and PAN were 0.02, 0.50, and 0.01 ppm, respectively, the actual ozone concentration, corrected for interference, would be

$$\text{Corrected } O_3 = (0.15 \text{ ppm } O_3) + (0.02 \text{ ppm } SO_2) \left(\frac{1 \text{ ppm } O_3}{\text{ppm } SO_2} \right)$$

$$- (0.50 \text{ ppm } NO_2) \left(\frac{0.1 \text{ ppm } O_3}{\text{ppm } NO_2} \right)$$

$$- (0.01 \text{ ppm PAN}) \left(\frac{0.5 \text{ ppm } O_3}{\text{ppm PAN}} \right)$$

$$= 0.115 \text{ ppm} \tag{5.8}$$

For this example, interferences result in a ozone measurement 20% above the actual value. If, on the other hand, this method were being used to measure ozone in a calibration gas mixture where the concentrations of NO_2, SO_2, and PAN were known to be very much lower than the ozone concentration, correction for interference would not be necessary.

A review of the literature on the performance of an analytical method is also useful. The Griess–Saltzman method is widely accepted for the determination of NO_2 (APHA, 1977). Under most ordinary conditions, interferences from SO_2, O_3, PAN, and other nitrogen oxides cause negligible errors (APHA, 1977). In a study of tobacco smoke chemistry, however, the Griess–Saltzman method was found to be 40% lower than the chemiluminescent procedure when measuring total NO_x per cigarette in mainstream tobacco smoke (Jenkins and Gill, 1980). The authors believe that part of the difference may be from a chemical interference, and do not recommend the Griess–Saltzman method for use in tobacco smoke polluted atmospheres.

Many factors contribute to the selection of an analytical method. For some applications, standard methods may be too expensive or not portable and non-standard methods may have to be used. Under these circumstances, a preliminary study of interferences should be made before starting an extensive field measurement program. This can be done as follows. Select a standard method with known characteristics to be used as a reference method in a comparison study. Using the standard method and the test method under evaluation, sample a calibration gas mixture and adjust or calibrate the test method's response to agree with the standard method's value. Both methods should now agree with the concentration of the calibration gas. Next, simultaneously collect a statistically significant number of samples in a typical environment to be studied with the method of interest and the standard reference method. If no significant differences between the methods are found, the test method can be used based on the laboratory calibration. If consistent differences are found, a correction factor or ratio of the standard method concentration divided by the test method concentration can be calculated and used to correct further measurements by the test method. This correction is only valid if the error is always in the same direction and magnitude as a percent of the standard method's value. If significant differences of varying magnitude are found, a laboratory study measuring calibration gas mixtures with fixed quantities of potentially interfering species, known to be present in the environments being studied, must be performed to quantitate the magnitude of the interference. Then, by measuring or estimating the concentration of interfering species, concentrations measured by the test method can be corrected to the actual concentration. This procedure can be quite time consuming and must be considered when planning a field measurement program.

Another approach to this problem is to remove the interference before it reacts to cause an error in a measurement. This can be done by selectively removing the interfering chemical species before collection or removing the effect of the collected species by distillation, ion-exchange resins, complexing agents, pH adjustment, and a variety of other techniques (APHA, 1977). Caution must be exercised when applying these techniques because they may have a significant effect on the pollutant of interest as well as removing the interfering species.

Interfering species may be present at the sampling site during sample collection, developed during sample storage, or imparted or developed in laboratory analysis. Consequently, it often is necessary to consider these steps, as well as the potential effects of chemical interferences in an environment being studied, in explaining differences between an untested method and a reference method.

Both wet chemical and instrumental methods can experience interferences. Mercury (Allen, 1977) and hydrocarbons contained in floor cleaning compounds (Selway, 1979) have been found to interfere with uv absorption techniques for ozone monitoring. Alcohol (Cuddeback et al., 1976; First and Hinds, 1976) and nitrogen oxides (Bay et al., 1972) are known to interfere with the electrochemical measurement of CO. Cigarette mainstream and sidestream smoke components evidently alter the chemiluminescent measurement of NO_x (Elam, 1981).

EFFECTS OF HUMAN ACTIVITIES

An indoor air characterization program should be designed to collect representative samples from the areas of concern. Human activity patterns are an important consideration. As an example, in a six-month study of indoor air pollution, total suspended particulate concentration was measured in two intensive care units in an urban hospital (Neal et al., 1978). The average indoor–outdoor ratio was 0.45. Unlike many other studies, indoor concentrations were not correlated with simultaneously measured outdoor levels. Indoor levels were a function of floor type, patterns of smoking, people movement, and ventilation practices— pointing out the importance of knowing normal activity patterns when interpreting the results of monitoring.

People also affect indoor pollutant concentrations by using pollution causing products and by emitting pollutants themselves. Emissions from pressurized consumer products (Table 4.15) and humans (Table 4.16) can affect indoor air quality.

PERSONAL EXPOSURES

Personal exposure is a function of the actual inhaled concentration of indoor pollution. In order to be certain that air samples are representative of personal exposure, it is necessary to collect samples as near as practical to the exposed person's breathing zone. General environmental samples of room air (referred to as area samples) may not give an accurate quantitation of personal exposure. This is especially true when the sources of pollution are near exposed people and room air mixing is not complete. In many instances, adequate personal sampling methods are not available, or it is not practical for exposed persons to carry the sampling equipment with them. Under these circumstances, area sampling techniques must be used. In general, area techniques are more sensitive than personal methods because a greater volume of air is sampled.

In a study of personal exposure to respirable particles and sulfates in homes, high correlations were found between personal exposure and indoor concentration (r^2 = 0.728 and 0.514 for sulfate and particles, respectively) (Dockery and Spengler, 1981). In addition, simultaneous measurements of personal, indoor, and outdoor respirable particle concentrations in two cities indicated that 24-hr mean concentrations for personal and indoor samples did not differ significantly from each other. However, both exceeded the corresponding outdoor concentration by approximately 20 $\mu g/m^3$ (Spengler et al., 1981). Consequently, for pollutants with no major indoor sources (e.g., sulfates), indoor measurements will be very comparable to personal measurements, probably due to the high percentage of time spent indoors. Also, as is discussed in Chapter 11, indoor measurements at various locations within a residence rarely differ by more than a factor of 2.

Analytical methods based on passive collection systems are receiving increased attention. Because these methods do not require an air pump and associated battery pack, passive samplers are very small, lightweight devices that can be worn near an exposed person's breathing zone without any discomfort or inconvenience. If properly calibrated, passive samplers can greatly simplify many of the problems associated with measuring personal exposures. These samplers, however, are limited to the collection of gaseous pollutants (see section on *Sampling and Analysis–Gases*).

It is a good practice to do backup area sampling when using personal sampling methods. The measurements from area sampling can be used as a check on the performance of the personal sampling technique, and, through the use of indoor air quality models, to estimate source strength for the pollutants of concern.

REFERENCES*

ACGIH (1978). American Conference of Governmental Industrial Hygienists. *Air Sampling Instruments for Evaluation of Atmospheric Contaminants.* Cincinnati, Ohio.

*Reference to particular manufacturers' literature does not constitute endorsement of their products.

Aerosol Technology Committee (1970). American Industrial Hygiene Association Guide for respirable dust sampling. *Am. Ind. Hyg. Assoc. J.* **31**:133–137.

Allen, R. J. (1977). *Relationship between indoor and outdoor concentrations of carbon monoxide and ozone for an urban hospital.* Ph.D. thesis, University of Illinois at the Medical Center, Chicago.

Allen, R. J. and Wadden, R. A. (1977). Modification and operation of a size-selective atmospheric particle sampler. *Atmos. Environ.* **11**:1101–1106.

Amass, C. E. (1979). Passive membrane-limited dosimeters using specific ion electrode analysis. *Proceedings of the symposium on the development and usage of personal monitors for exposure and health effect studies,* U.S. Environmental Protection Agency, EPA-600/9-79-032, June.

American Gas and Chemical Co. (1981). Leak-Tec Personal Protection Indicators Northvale, New Jersey.

APHA (1975). *Standard Methods for the Examination of Water and Wastewater,* 14th ed. American Public Health Association, Washington, D.C.

APHA (1977). *Methods of Air Sampling and Analysis.* APHA Intersociety Committee, American Public Health Association, Washington, D.C.

Bamberger, R. L., Esposito, G. G., Jacobs, B. W., Podalak, G. E., and Mazur, J. F. (1978). A new personal sampler for organic vapors. *Am. Ind. Hyg. Assoc. J.* **39**:701–708.

Bay, W. H., Blureton, K. F., Lieb, H. C., and Oswin, H. G. (1972). Electrochemical measurement of carbon monoxide. *American Laboratory,* July.

Bell, D. R., Reiszner, K. D., and West, P. W. (1975). Permeation method for the determination of average concentration of carbon monoxide in the atmosphere. *Anal. Chim. Acta.* **77**:245–254.

Charlson, R. J., Ahlquist, N. C., Selvidge, H., and MacCready, P. B. (1969). Monitoring of atmospheric aerosol parameters with the integrating nephelometer. *J. Air Pollut. Control Assoc.* **12**:937–942.

Cuddeback, J. E., Donovan, J. R., and Burg, W. R. (1976). Occupational aspects of passive smoking. *Am. Ind. Hyg. Assoc. J.* **37**:263–267.

Dockery, D. W. and Spengler, J. D. (1981). Indoor–outdoor relationship of respirable sulfates and particles. *Atmos. Environ.* **15**:335–343.

Elam, L. (1981). *Passive and active measurement of nitrogen oxides in cigarette smoke.* M.S. Thesis, School of Public Health, University of Illinois, Chicago.

El-Shobokshy, M. S. and Ismail, I. A. (1980). Deposition of aerosol particle from turbulent flow onto rough pipe wall. *Atmos. Environ.* **14**:297–304.

E.I. DuPont (1981). High Quality Air and Noise Monitoring Instruments, DuPont Occupational and Environmental Health Products, Applied Technology Division, Wilmington, D.C.

EPA (1971). Environmental Protection Agency. *Fed. Reg.* **36**:22388–22392.

EPA (1978a). *Survey on research needs on personal samplers for toxic organic compounds.* U.S. Environmental Protection Agency, EPA-600/8-78-004, April.

EPA (1978b). *Air quality criteria for ozone and other photochemical oxidants.* U.S. Environmental Protection Agency, EPA-600/8-78-004, April.

EPA (1981). Workshop on indoor air quality research needs, Environmental Protection Agency, Department of Energy, Interagency research group on indoor air quality, Leesburg, Virginia, April.

Evans, M., Molyneux, M., Sharp. T., Bailey, A., and Hollingdale-Smith, P. (1977). The practical application of the Porton diffusion sampler for the measurement of time weighed average exposure to volatile organic substances in air, *Ann. Occup. Hyg.* **20**:357–363.

First, M. W. and Hinds, W. C. (1976). Ambient tobacco smoke measurement. *Am. Ind. Hyg. Assoc. J.* **37**:655–657.

Gillespie, J. C. and Daniel, L. B. (1979). A new sampling tool for monitoring exposures to toxic gases and vapors. *Proceedings of the symposium on the development and usage of personal monitors for exposure and health effect studies*, U.S. Environmental Protection Agency, EPA-600/9-79-32, June.

Hardy, J. K., Dasgupta, P. K., Reiszner, K. D., and West, P. W. (1979). A personal chlorine monitor utilizing permeation sampling, *Environ. Sci. Technol.* **13**:1090–1093.

Hardy, J. K., Strecker, D. T., Savariar, C. P., and West, P. W. (1981). A method for the personal monitoring of hydrogen sulfide utilizing permeation sampling, *Am. Ind. Hyg. Assoc. J.* **42**:781–786.

HEW (1979). *Smoking and Health—A Report of the Surgeon General.* U.S. Department of Health Education and Welfare. Pub. No. (PHS)79-50066, January.

Hickey, J. L. S. and Bishop, C. C. (1981). Field comparison of charcoal tubes and passive vapor monitors with mixed organic vapors. *Am. Ind. Hyg. Assoc. J.* **42**:264–267.

Hill, R. H. and Fraser, D. A. (1980). Passive dosimetry using detector tubes. *Am. Ind. Hyg. Assoc. J.* **41**:721–729.

Jenkins, R. A. and Gill, B. E. (1980). Determination of oxides of nitrogen (NO_x) in cigarette smoke by chemiluminescent analysis. *Anal. Chem.* **52**:925–928.

John, W. and Reischl, G. (1978). Measurements of the filtration efficiencies of selected filter types. *Atmos. Environ.* **12**:2015–2019.

Kim, C. S., McDonald, R., and Sackner, M. A. (1981). Generation and characterization of sulfate aerosols for laboratory studies. *Am. Ind. Hyg. Assoc. J.* **42**:521–528.

Lautenberger, W. J., Kring, E. V., and Morello, J. A. (1980). A new personal badge monitor for organic vapors. *Am. Ind. Hyg. Assoc. J.* **41**:737–747.

Liu, B. Y. H. and Lee, K. W. (1976). Efficiency of membrane and nucleopore filters for submicrometer aerosols. *Environ. Sci. Technol.* **10**:345–350.

Liu, B. Y. H. and Pui, D. Y. H. (1975). On the performance of the electrical aerosol analyzer. *J. Aerosol Sci.* **6**:249–264.

Liu, B. Y. H. and Pui, D. Y. H. (1974). A submicron aerosol standard and the primary, absolute calibration of the condensation nuclei counter. *J. Colloid. Inter. Sci.* **47**:155–171.

Liu, B. Y. H., Pui, D. Y. H., Rubow, K. L., and Kuhlmey, G. A. (1978). *Research on air sampling filter media.* University of Minnesota Particle Technology Laboratory Progress Report EPA Grant R804600, Minneapolis, May.

Lundgren, D. A., Carter, L. D., and Daley, P. S. (1976). Aerosol mass measurement using piezoelectric crystal sensors. In: *Fine Particles: Aerosol Generation, Measurement, Sampling and Analysis.* B. Y. H. Liu, Ed., Academic Press, New York.

Marple, V. and Willeke, K. (1976). Impactor design. *Atmos. Environ.* **10**:891–896.

MDA (1981). The monitox personal gas detection alarm system, Environmental division, MDA Scientific, Inc., Glenview, Ill.

Miller, F. J., Gardner, D. E., Graham, J. A., Lee, R. E., Wilson, W. E., and Bachmann, J. D. (1979). Size considerations for establishing a standard for inhalable particles. *J. Air Pollut. Control Assoc.* **29**:610–615.

Monitor Labs, Inc. (1975). Instrument Manual, Oxides of Nitrogen Analyzer Model 8440.

Neal, A. deW., Wadden, R. A., and Rosenberg, S. H. (1978). Evaluation of indoor particulate concentrations for an urban hospital. *Am. Ind. Hyg. Assoc. J.* **39**:578–582.

Nelms, L. H., Reizner, K. D., and West, P. W. (1977). Personal vinyl chloride monitoring device with permeation technique for sampling. *Anal. Chem.* **49**:994–997.

Neustadter, H. E., Sidik, S. M., King, R. B., Fordyce, J. S., and Burr, J. C. (1975). The use of Whatman-41 filters for high-volume air sampling. *Atmos. Environ.* **9**:101–109.

NIOSH (1973). *The Industrial Environment—Its Evaluation and Control.* National Institute for Occupational Safety and Health, U.S. Department of Health Education and Welfare-Public Health Service, Cincinnatti.

NIOSH (1977). *Manual of Analytical Methods,* Vols. I–III. National Institute for Occupational Safety and Health, Washington, D.C.

Obermayer, A. S., Nichols, L. D., and Gould, A. S. (1980). Improved colorimetric passive dosimeters. Presented at the National Meeting of the American Industrial Hygiene Association, May.

Orofino, T. A. and Usmani, A. M. (1980). Passive dosimetry. *Am. Lab.,* pp. 96–104, July.

Palmes, E. D. and Gunnison, A. F. (1973). Personal monitoring device for gaseous contaminants. *Am. Ind. Hyg. Assoc. J.* **34**:78–81.

Palmes, E. D. and Tomczyk, C. (1979). Personal sampler for NO_x. *Am. Ind. Hyg. Assoc. J.* **40**:588–591.

Palmes, E. D., Gunnison, A. F., DiMattio, J., and Tomczyk, D. (1976). Personal sampler for nitrogen dioxide. *Am. Ind. Hyg. Assoc. J.* **37**:570–577.

Raabe, O. G. (1976). The generation of aerosols of fine particles. In: *Fine Particles: Aerosol Generation, Measurement, Sampling and Analysis.* B. Y. H. Liu, Ed., Academic Press, New York.

Reid, R. C. and Sherwood, T. K. (1966). *The Properties of Gases and Liquids.* 2nd ed., McGraw-Hill, New York, 646 pp.

Reiszner, K. D. and West, P. W. (1973). Collection and determination of sulfur dioxide incorporating permeation and West-Gaeke procedure. *Environ. Sci. Technol.* 7:526–532.

Repace, J. L. and Lowrey, A. H. (1980). Indoor air pollution, tobacco smoke, and public health. *Science* 208:464–472.

Scheff, P. A. and Wadden, R. A. (1979). Comparison of three methods of particulate measurement in Chicago air. *Atmos. Environ.* 13:639–643.

Selway, M. D. (1979). *Characterization of photocopying machines as an indoor source of ozone.* M.S. Thesis, School of Public Health, University of Illinois, Chicago.

SKC (1981). Catalog and guide to air sampling standards, SKC Inc., Eighty Four, PA.

Spengler, J. D., Treitman, R. D., Tosteson, T. D., and Mage, D. T. (1981). Personal exposures to respirable particulates: A tale of two cities—Kingston and Harriman, Tennessee. Harvard University School of Public Health, Boston, Massachusetts. International Symposium on Indoor Air Pollution, Health and Energy Conservation, University of Massachusetts, Amherst, Massachusetts, October.

Stern, A. C. (1976). *Air Pollution, Volume III. Measuring, Monitoring and Surveillance of Air Pollution.* Academic Press, New York.

Tompkins, F. C. and Goldsmith, R. L. (1977). A new personal dosimeter for the monitoring of industrial pollutants. *Am. Ind. Hyg. Assoc. J.* 38:371–377.

Waggoner, A. P. and Charlson, R. J. (1977). *Aerosol Characteristics and Visibility,* U.S. Environmental Protection Agency-PB-269-944.

Wallace, L. A. and Ott, W. R. (1982). Personal Monitors: A State-of-the-Art Survey. *J. Air Pollut. Control Assoc.* 32:601–610.

West, P. W. and Reiszner, K. D. (1978). Field tests of a permeation-type personal monitor for vinyl chloride. *Am. Ind. Hyg. Assoc. J.* 39:645–650.

West, P. W. and Reiszner, K. D. (1979). Personal monitoring by means of gas permeation. *Proceedings of the symposium on the development and usage of personal monitors for exposure and health effects studies,* U.S. Environmental Protection Agency, EPA-600/9-79-032, June.

West, P. W. (1980). Passive monitoring of personal exposures to gaseous toxins. *Am. Lab.,* pp. 35–39, July.

Willeke, K. (1975). Performance of the slotted impactor. *Am. Ind. Hyg. Assoc. J.* 36:683–691.

Willeke, F. and Whitby, K. T. (1975). Atmospheric Aerosols: Size Distribution Interpretation. *J. Air Pollut. Control Assoc.* 25:529–534.

PART TWO

INDOOR AIR QUALITY PREDICTION

6

AIR QUALITY MODELS

Several approaches have been used to estimate expected indoor pollutant concentrations. These include deterministic models based on a pollutant mass balance around a particular indoor volume; a variety of empirical approaches based on statistical evaluation of test data and usually least-squares regression analysis; and a combination of both forms, empirically fitting the parameters of the mass balance with values statistically derived from experimental measurements. Each approach has advantages. The mass balance model provides more generality in application. But empirical models, for application within the range of measurements from which they were developed, may provide much more accurate information.

ONE-COMPARTMENT MODELS

The mass balance for pollutant flow into and out of an indoor volume, including recycling and interior sources and sinks, is described in Figure 6.1 and expressed by

Air mass balance: $\quad q_0 + q_2 = q_3 + q_4$ $\hfill$ (6.1)

Pollutant mass balance: $\quad V\dfrac{dC_i}{dt} = kq_0 C_o (1 - F_0) + kq_1 C_i (1 - F_1)$

$$+ kq_2 C_o - k(q_0 + q_1 + q_2)C_i + S - R \qquad (6.2)$$

where C is concentration indoors (C_i) and outdoors (C_o); t is time; q is volumetric flow rate for make-up air (q_0), recirculation (q_1), infiltration (q_2), exfiltration (q_3), and exhaust (q_4); F is filter efficiency for make-up (F_0) and recirculation air (F_1) (often the same); V is room volume; S is indoor source emission rate; R is indoor sink removal rate; and k, a factor which accounts for inefficiency of mixing, is the fraction of incoming air which completely mixes within the room volume. The solution of Equation (6.2) for the change in C_i with t, hold-

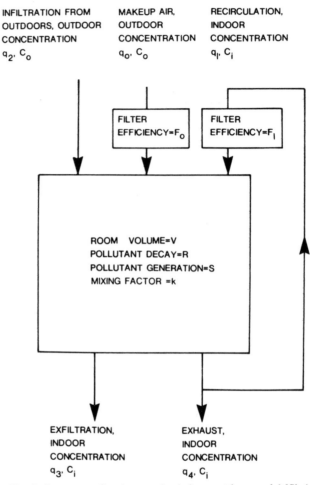

INFILTRATION FROM
OUTDOORS, OUTDOOR
CONCENTRATION
q_2, C_o

MAKEUP AIR,
OUTDOOR
CONCENTRATION
q_0, C_o

RECIRCULATION,
INDOOR
CONCENTRATION
q_1, C_i

FILTER
EFFICIENCY=F_o

FILTER
EFFICIENCY=F_1

ROOM   VOLUME=V
POLLUTANT DECAY=R
POLLUTANT GENERATION=S
MIXING FACTOR =k

EXFILTRATION,
INDOOR
CONCENTRATION
q_3, C_i

EXHAUST,
INDOOR
CONCENTRATION
q_4, C_i

Figure 6.1. Ventilation system for time varying indoor–outdoor model [Shair and Heitner (1974). Reprinted by permission of the American Chemical Society from *Environmental Science and Technology*].

ing all other factors constant and with boundary values $C_i = C_s$ at $t = 0$, is

$$C_i = \frac{k[q_0(1 - F_0) + q_2]C_o + S - R}{k(q_0 + q_1F_1 + q_2)} [1 - e^{-(k/V)(q_0+q_1F_1+q_2)t}]$$
$$+ C_s e^{-(k/V)(q_0+q_1F_1+q_2)t} \tag{6.3}$$

For the case where R is a first-order function of C_i, the solution will have the form

$$C_i = \frac{k[q_0(1-F_0)+q_2]C_o+S}{k(q_0+q_1F_1+q_2)+E}\{1-e^{-[(k/V)(q_0+q_1F_1+q_2)+E]t}\}$$

$$+ C_s e^{-[(k/V)(q_0+q_1F_1+q_2)+E]t} \tag{6.4}$$

where E is a proportionality constant for the particular pollutant of interest, such that $R = EC_i$. Steady-state values of indoor concentrations, $C_{i,ss}$, will result by letting t approach ∞ which, for Equation (6.4), results in

$$C_{i,ss} = \frac{k[q_0(1-F_0)+q_2]C_o+S}{k(q_0+q_1F_1+q_2)+E} \tag{6.5}$$

Equation (6.3), or similar forms, have been applied in analyzing indoor ozone decay (Shair and Heitner, 1974; Sabersky et al., 1973; Allen et al., 1978; Selway et al., 1980), odor control (Turk, 1963), particulate matter and carbon monoxide (Cote and Holcombe, 1971), ozone and carbon monoxide (Allen and Wadden, 1982), particulate matter from cigarette smoke (Hoegg, 1972; Ishizu, 1980), CO_2 from respiration (Kusuda, 1976), a variety of pollutants in residential settings (Moschandreas et al., 1978), and energy efficient control strategies (Woods et al., 1981). A similar approach involving activity balances for Ra-222 and radon daughters was described by Equations (4.1) and (4.2).

INFILTRATION ESTIMATION

It is apparent that many factors are required to use Equation (6.3). Infiltration, q_2, and exfiltration, q_3, are a function of the temperature and pressure differences between indoor and outdoor air. Temperature differences cause the stack effect, which is a consequence of warm air rising. Static pressure differences result from changes in wind speed. A simplified statistical model which reflects, to some extent, these two regimes is

$$\frac{q_2}{V} \simeq \frac{q_3}{V} = 0.315 + 0.0273\,v + 0.0105|T_A - T_R| \tag{6.6}$$

where v is wind speed in miles/hr; q_2/V is in hr^{-1}; and T_A, the ambient temperature, and T_R, the room temperature, are both in $°F$ (Achenbach and Coblentz, 1963). The coefficients have been multiplied by 1.25 to simulate the existing housing stock (Roseme et al., 1979). Equation (6.6) can be used as a first estimate of q_2, particularly in the absence of forced make-up or recirculation air convection.

Another simplified approach for residences is to use an average number of air changes per hour appropriate for particular applications. Table 6.1 indicates typical values (ASHRAE, 1981).

It is also possible to estimate infiltration rates more specifically as functions of crack area and pressure drop. The static pressure, P_v, over a building surface

Table 6.1   Air Changes Occurring Under Average Conditions in Residences, Exclusive of Air Provided for Ventilation[a]

Type of Room	Air Changes per Hour[b]
No windows or exterior doors	0.5
Windows or exterior doors on one side	1
Windows or exterior doors on two sides	1.5
Windows or exterior doors on three sides	2
Entrance halls	2

[a] ASHRAE (1981).

[b] For rooms with weather-stripped windows or with storm sash, use $\frac{2}{3}$ of these values. It has also been found that since air flows in half of the openings and out the other half, the values should be multiplied by 0.5 (Janssen et al., 1980).

can be described by

$$P_v = 0.6008 v^2 \tag{6.7}$$

where v is the wind velocity in m/s and P_v is in Pa (1 in. water = 249.1 Pa) (ASHRAE, 1981). The pressure difference due to a thermal gradient, ΔP_c, will be given by

$$\Delta P_c = 0.0342 Ph \left(\frac{1}{T_o} - \frac{1}{T_i} \right) \tag{6.8}$$

where ΔP_c is in Pa; P is atmospheric pressure, Pa; T_o is outside temperature, °K; T_i is inside temperature, °K; and h is the distance from neutral pressure level or effective chimney height, m (ASHRAE, 1981). If the cracks and openings are uniformly distributed in the vertical direction, h will be one half the building height. The total pressure drop across the wall on the windward side will then be $P_v + \Delta P_c$.

The flow resulting from these pressure differences is expressed as

$$q_2 \simeq q_3 = K_F (\Delta P)^n \tag{6.9}$$

where ΔP is the pressure between indoors and outdoors, K_F is a flow coefficient, and n is an empirical exponent between 0.5 (turbulent flow) and 1.0 (laminar flow) and is usually chosen as 0.65. Also,

$$\Delta P = \frac{P_v + \Delta P_c}{1 + (A_W/A_L)^{1/n}} \tag{6.10}$$

where A_W and A_L are the leakage areas on the windward and leeward sides. For a square house with the wind direction normal to one of the sides, Equation (6.10) becomes $\Delta P = 0.85(P_v + \Delta P_c)$ when $n = 0.65$.

In practice, Equation (6.9) is usually not applied directly because of the wide variation in the values of K_F. For many applications infiltration rates have been determined as a function of pressure drop in field and laboratory settings, and these may be used for design purposes. Tables 6.2, 6.3, and 6.4 show infiltration rates for various types of openings (Janssen et al., 1980; ASHRAE, 1981). The values of q_2 (or q_3) can be estimated by determining the crack length for each window sash and door perimeter (sash leakage due to clearance needed in order to open doors and windows), the perimeter of each window frame and door (frame leakage due to cracks between frame and wall), the wall area minus the areas of windows and doors (Table 6.3), and using the infiltration rates from Tables 6.2–6.4. Typically, calculations will be made at 75 Pa since this is the only pressure difference specified in Table 6.4. The flow rate at the actual pressure differential in Pa $[P_v + \Delta P_c$ calculated from Equations (6.7) and (6.8)] can be determined from a form of Equation (6.9):

$$q_{2,\Delta P} = q_{2,\,75\ \text{Pa}} \left(\frac{\Delta P}{75}\right)^{0.65} \tag{6.11}$$

Table 6.2   Infiltration Through Double-Hung Wood Windows[a]

	Pressure Difference (Pa)				
Type of Window	25	50	75	100	125
A. Wood double-hung window (locked)					
1. Nonweather-stripped loose fit[b]	2.0	3.1	3.9	5.0	5.8
2. Nonweather-stripped, average fit, or weather-stripped, loose fit[c]	0.70	1.1	1.5	1.8	2.1
3. Weather-stripped, average fit	0.36	0.59	0.77	0.93	1.1
B. Frame-wall leakage[d] (leakage is that passing between the frame of a wood double-hung window and the wall)					
1. Around frame in masonry wall, not caulked	0.43	0.67	0.88	1.1	1.2
2. Around frame in masonry wall, caulked	0.08	0.13	0.15	0.18	0.21
3. Around frame in wood frame wall	0.34	0.54	0.75	0.90	1.1

[a] Measured in liters/s per meter of crack (ASHRAE, 1981; Janssen et al., 1980).
[b] A 2.4-mm crack and clearance represent a poorly fitted window, much poorer than average.
[c] The fit of the average double-hung wood window was determined as 1.6-mm crack and 1.2-mm clearance by measurements on approximately 600 windows under heating season conditions.
[d] The values given for frame leakage are per meter of sash perimeter, as determined for double-hung wood windows. Some of the frame leakage in masonry walls originates in the brick wall itself and cannot be prevented by caulking. For the additional reason that caulking is not done perfectly and deteriorates with time, it is considered advisable to choose the masonry frame leakage values for caulked frames as the average determined by the caulked and noncaulked tests.

Table 6.3   Infiltration Through Walls[a]

Type of Wall	Pressure Difference (Pa)				
	12	25	50	75	100
Brick wall[b]					
8.5 in. Plain	0.42	0.76	1.35	2.03	2.37
Plastered[c]	0.004	0.007	0.012	0.017	0.023
13 in. Plain	0.42	0.68	1.19	1.69	2.03
Plastered[c]	0.001	0.003	0.004	0.008	0.009
Plastered[d]	0.003	0.020	0.039	0.056	0.071
Frame wall					
Lath and plaster[e]	0.008	0.013	0.019	0.025	0.027
Shingles					
16-in. Shingles on 1 X 4 in. boards on 5 in. centers with paper; or 16-in. shingles, shiplap and paper	0.021	0.043	0.085	0.13	0.17
18-in. Shingles on shiplap	0.76	1.2	2.2	2.9	3.6
24-in. Shingles on shiplap	1.9	3.5	5.8	7.6	9.4
16-in. Shingles on 1 X 4 in. boards on 5 in. centers	3.4	5.7	8.9	11.4	14.0
24-in. Shingles on 1 X 6 in. boards on 11 in. centers	6.0	9.9	15.7	20.3	24.9

[a]Leakage in liters/sec per square meter of wall area (ASHRAE, 1981).
[b]Constructed of porous brick and lime mortar; workmanship poor.
[c]Two coats prepared gypsum plaster on brick.
[d]Furring, lath, and two coats prepared gypsum plaster on brick.
[e]Wall construction: bevel siding painted or cedar shingles, sheathing, building paper, wood lath, and three coats gypsum plaster.

The value of $P_v + \Delta P_c$ can be adjusted to ΔP with Equation (6.10), but Equation (6.11) can also be used with the ratio $(P_v + \Delta P_c)/75$.

The infiltration values given in Tables 6.2 and 6.4 are in essential agreement with actual measurements. For instance, average leakage values for wood double-hung windows without storm windows (Table 6.2) have been reported to vary from 0.93 to 3.3 liters/sec $\cdot$ m^2 at 75 Pa (Tamura, 1975). However, actual wall leakage rates at 75 Pa have been reported from 3.4 to 6.2 liters/sec $\cdot$ m^2 through frame construction and brick veneer or metal siding; 0.76–1.02 liter/sec $\cdot$ m^2 for stucco finish; and 1.37–4.82 liters/sec $\cdot$ m^2 for ceilings (Tamura, 1975). These field measurement values are much larger than those in Table 6.3 which were developed from laboratory tests of wall sections (ASHRAE, 1981).

Table 6.4 Window and Door Specification[a]

Specification/Material	Type of Class	Air Leakage[b]
ANSI A134.1	A-B1 (Awning)	1.16
Aluminum	A-A2 (Awning)	0.77
	C-B1, C-A2, C-A3 (Casement)	0.77
	DH-B1 (Hung)	1.16
	DH-A2, DH-A3, DH-A4 (Hung)	0.77
	HS-B1, HS-B2, HS-A2 (Sliding)	1.16
	HS-A3 (Sliding)	0.77
	J-B1 (Jalousie)	7.62 liter/sec · m^2
	JA-B1 (Jai-Awning)	1.16
	P-B1, P-A2 (Projected)	0.77
	P-A2, 50 (Projected)	0.58
	P-A3 (Projected)	0.77[c]
	TH-A2 (Inswinging)	0.58
	TH-A3 (Inswinging)	0.77[c]
	VP-A2 (Pivoted)	0.58
	VP-A3 (Pivoted)	0.77[c]
	VS-B1 (Vertical sliding)	1.16
ANSI A134.2	SGD-B1 (Sliding glass door)	5.08 liter/sec · m^2
Aluminum	SGD-B2, SGD-A2, (Sliding glass door)	2.54 liter/sec · m^2
	SGD-A3 (Sliding glass door)	2.54 liter/sec · m^{2c}
ANSI A200.1	All types windows	
Wood	Class A	0.77
	Class B	0.77
ANSI A200.2	All types sliding glass doors	2.54 liter/sec · m^2
Wood		
Fed. MHC & SS[d]	Windows (all types)	2.54 liter/sec · m^2
280.403	Sliding glass doors	
Fed. MHC & SS[d]	Vertical Entrance	5.08 liter/sec · m^2
280.405		

[a] Leakage in liters/sec per meter of crack unless noted otherwise (ASHRAE, 1981; Janssen et al., 1980).
[b] At 75 Pa (0.30 in. water) pressure or 11.2 m/s (25 mi/hr) wind velocity.
[c] At 300 Pa (1.30 in. water) pressure or 22.3 m/s (50 mi/hr) wind velocity.
[d] Federal Mobile Home Construction and Safety Standard.

The Tamura study also found for six Canadian houses an overall leakage area [the equivalent orifice area determined at 75 Pa and $n = 0.5$ in Equation (6.9)] of 0.63×10^{-4} ft^2/ft^3 house volume. Outside wall leakage area exclusive of windows and doors, but including leakage between wall and door and window frames, ranged from 0.016 to 0.130 $in.^2/ft^2$ wall area and 0.03 to 0.1 $in.^2/ft^2$ ceiling area. Leakage values through stucco finish walls were only 10–30% of those through masonry, or a combination of masonry and aluminum siding or asbestos shingles. A comparison of the two ASHRAE methods with actual tracer measurements (using methane) has been carried out by Janssen et al. (1980) for four houses. Table 6.5 summarizes their findings along with estimates based on Equation (6.6). They also determined that ΔP_c for short buildings was not a large contribution to the total pressure difference. The estimated values are not unreasonable given the simplicity of the models.

While windows and doors have been identified as major infiltration sites, other openings may also be important. Measurements on Canadian houses indicated that windows and doors accounted for only 15–24% of total leakage at 75 Pa (Tamura, 1975). In a test of 50 Texas houses (no basements; 19% of the air conditioning duct systems in air conditioned space), the air flow through electrical wall outlets was estimated at 20% of total infiltration (Caffey, 1979). Another major leakage point was the soleplate (the bottom plate or baseboard area of the exterior wall) which allowed 25% of the infiltration. Table 6.6 lists typical leakage rates measured at 62.25 Pa (0.25 in H_2O). Actual infiltration rates would be calculated by dividing the total leakage value by 4 (infiltration only across ∼1 wall), determining the actual pressure drop from atmospheric conditions using Equations (6.7) and/or (6.8), and adjusting to actual pressure difference using Equation (6.11). A soleplate contribution to infiltration roughly twice that of window and doors has also been reported by Cole et al., (1980).

While 75 Pa is a useful test point for determining leakage locations, actual ΔP values are probably between 1 and 5 Pa (0.004–0.02 in. H_2O). In one study, the average ΔP for 13 houses and 63 measurement periods was 1.2 Pa with the standard deviation $\sigma = 1.6$ (Grimsrud et al., 1979). Using a reference ΔP of 4 Pa and literature data, the mean calculated infiltration rate for 224 North American houses during the heating season was 0.67 air change/hr (ach) with $\sigma = 0.48$ ach (Grimsrud et al., 1980; Sherman and Grimsrud, 1980a). The mean value for 111 houses built within 2 years of the study date was 0.48 ach with $\sigma = 0.24$. "Tight" housing may have infiltration rates more typically varying between 0.2 and 0.5 air changes/hour.

A study of low-income housing, 10–90 years old, in 14 cities in all climatic zones of the United States, indicated mean infiltration values between 0.5 and 1 ach (Grot and Clark, 1979). However, 20% of the houses had infiltration rates exceeding 1.5 ach. Typical histograms of these infiltration measurements are given in Figure 6.2.

Table 6.5 Comparison of ASHRAE Methods to Calculate Infiltration[a]

| House[b] | Measured Windspeed (m/s) | Furnace Condition | Measured Infiltration ach | Calculated Infiltration, q_2/V | | | | | | |
| | | | | Air Change[c] | | Crack[d] | | | General[e] |
				Gross	Adjusted	11.2 m/s	6.7 m/s	2.2 m/s	At Measured Windspeed
MED II, So. Calif.	2–4	Sealed	0.3	1.21	0.60	1.59	0.82	0.20	0.55–0.57
Walnut Creek, Calif.	4	On	0.75	1.09	0.55	1.58	0.82	0.20	
Walnut Creek, Calif.		Off	0.13						0.38
Minnetonka, Minn.	4	On	0.49	0.30	0.40	0.84	0.43	0.10	
Minnetonka, Minn.	4	Off	0.46						0.89
New Brighton, Minn.	4–6	On	0.50	0.77	0.39	0.94	0.49	0.12	0.80–0.82

[a] Janssen et al., 1980.
[b] House volumes: MED II, 245 m³; Walnut Creek, 230 m³; Minnetonka, 690 m³; New Brighton, 690 m³.
[c] Table 6.1. Adjusted values are one half the values from Table 6.1 with weather-stripping taken into account for both gross and adjusted values.
[d] Equations (6.7), (6.8), (6.9), and (6.11); Tables 6.2–6.4.
[e] Equation (6.6).

Table 6.6   House Infiltration Sites[a]

Leakage Site	Number of Items	Leakage Rate (ft^3/min)	% of Total Leakage
Soleplate	175 ft of linear crack	630	24.6
Electrical wall outlet	65	520	20.3
Air conditioning duct system	1	345	13.5
Exterior window	13	300	11.8
Fireplace (damper closed)	1	139	5.5
Range vent (dampered)	1	132	5.2
Recessed spot light	4	132	5.2
Exterior door	3	117	4.6
Dryer vent	1	71	2.8
Sliding glass door	1	43	1.7
Bath vent	1	33	1.3
Other (often due to poor seal at contact between a brick fireplace and interior wall)		96	3.5

[a] Average data from 50 Texas houses taken at $\Delta P = 62.25$ Pa (Caffey, 1979). Floor area, 1780 ft^2; total leakage, 2558 ft^3/min.

Infiltration calculation methods for larger buildings are given in ASHRAE (1981). Equations (6.7) and (6.8) are still applicable but the stack effect will be much more important. Figures 6.3 and 6.4 describe infiltration flow per m^2 of exterior wall for supermarkets and shopping malls, schools, and high-rise office buildings (Shaw, 1981).

Several other procedures of varying complexity are available for calculating infiltration rates (ASHRAE, 1981; Sherman and Grimsrud, 1980a,b; Shaw, 1980; Cole et al., 1980). Shielding due to fences, trees, topography, or other buildings has also been identified as a factor that reduces infiltration (Shaw, 1980; Blomsterberg and Harrje, 1979; Malik, 1978; Mattingly and Peters, 1977). However, the actual decrease in leakage is less well defined. In a Princeton University study of an occupied residence before, during, and after a temporary tree windbreak was installed, air infiltration was reduced from 1.13 to 0.66 ach for a windspeed of 5.6 m/s and air temperature of 18°C (NAS, 1981). It has also been observed that relatively low make-up air rates combined with high recirculation tend to increase infiltration, possibly because of changes in static pressure differences between indoors and outdoors (Shair and Heitner, 1974).

SOURCE AND SINK TERMS

Emission factors (S) for many common sources are given in the section on *Indoor Sources*. Filter efficiencies are discussed in two following sections, *Particle*

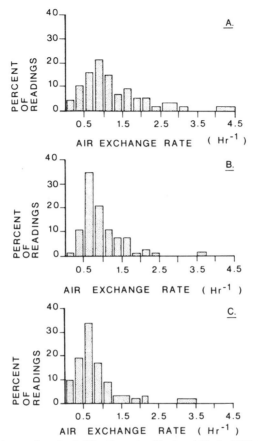

Figure 6.2. Low-income housing infiltration rates (Grot and Clark, 1979; NAS, 1981).

A. Charleston, S.C.; 134 measurements in 23 houses; average = 1.20 hr^{-1}; σ = 0.86 hr^{-1}.
B. Colorado Springs, Colorado; 114 measurements in 23 houses; average = 0.82 hr^{-1}; σ = 0.51 hr^{-1}.
C. Fargo, North Dakota; 83 measurements in 17 houses; average = 0.77 hr^{-1}; σ = 0.57 hr^{-1}.

Removal Devices and *Gas Filters and Traps*. The value of R is often unknown and is frequently incorporated with S which results in a net source term. However, some pollutants have been found to follow first-order decay rates. Ozone, in particular, tends to follow the form

$$R_{O_3} = \sum_{j=1} K_{dep_j} A_j C_i \qquad (6.12)$$

where K_{dep_j} is a deposition or decay velocity and A_j is an area of contact (Sabersky et al., 1973; Mueller et al., 1973). With reference to Equation (6.4), $E = K_{dep_j} A_j$. Ozone deposition velocities [rate of decay of O_3 (ng/cm^2 · min) per

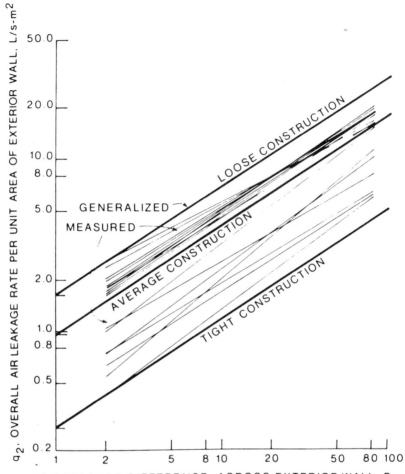

Figure 6.3. Generalized overall air leakage rates for supermarkets and shopping malls [Shaw (1981). Reprinted by permission from The *ASHRAE Journal*.]

mean concentration of O_3 in air (ng/cm^3)] are listed for several substances in Table 6.7 (Sabersky et al., 1973; Mueller et al., 1973; Sutton et al., 1976). The decay or deposition of other substances is quite possibly a function of the chemical and physical characteristics of room surfaces. For example, Table 6.8 summarizes data for SO_2 on carpets, wallpaper, and painted surfaces. Deposition velocities were greater for neutral or alkaline carpets than for those with acidic pH (Walsh et al., 1977). The removal term can also be expressed as a decay rate typically with the dimensions of inverse time. The decay rate, K, and deposition velocity are related by

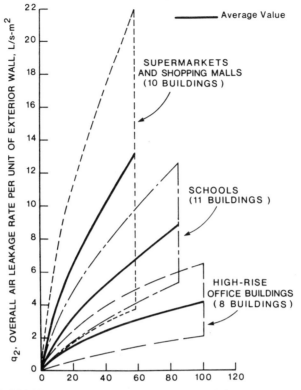

Figure 6.4. Comparison of overall air leakage for various building classifications [Shaw (1981). Reprinted by permission from The *ASHRAE Journal*.]

$$K_{dep} = K \frac{V}{A} \qquad (6.13)$$

A value of K_{NO_2} (cm/min) = (0.014 min^{-1}) $\times$ (V/A, cm) has been derived from the NO_2 decay data of Wade et al. (1975). (See Chapter 11 for derivation.)

MIXING FACTOR

The value of k, the mixing factor, is traditionally determined from the slope of a plot of the log of concentration versus time. To calculate k the slope is divided by q/V, the inverse of the time for one air change for a simple mass balance system. While it does represent a deviation from ideal mixing, its application to infiltration, recirculation, and make-up air probably does not exactly reflect the mixing process.

Table 6.7   Ozone Deposition Velocities for Various Surfaces and Indoor Areas

	Deposition Velocity K_{dep} (cm/min)	
	Sabersky et al. (1973)	Sutton et al. (1976)
Material		
Cotton muslin	0.88–6.52	0.366
Lamb's wool	0.24–6.34	5.21 (carpet)
Neoprene	0.91–5.79	
Plywood (1 side varnished)	0.30–1.83	
Nylon	0.03–1.92	11.9 (carpet)
Polyethylene sheet	0.61–1.46	0.091
Linen	0.33–0.56	
Lucite	0.03–0.37	
Aluminum	0.03–0.06	
Plate glass	0.03–0.06	
Latex paint		0.122
Enclosed Areas (Mueller et al., 1973)		
Aluminum room (12 m^3)	1.65	
Stainless steel room (15 m^3)	0.93	
Bedroom (14 m^3)	3.73	
Office (55 m^3)	2.23	
Ventilation Systems (Shair, 1981)		
23 ventilation systems in 17 laboratory-office buildings	1.22–4.88 (average = 2.13)	

The value of k has typically been estimated at 0.33–0.10. Drivas et al. (1972) using SF_6 as a tracer found that k varied from 0.30 to 0.60 for three relatively small rooms ($V = 41$ m^3) with simple in-and-out ventilation and no recirculation. Larger rooms would presumably be less well mixed. With good mixing (using fans) they found k values close to unity (Drivas et al., 1972; Shair and Heitner, 1974). Tests of cigarette smoke particles in rooms with $V = 16, 71, 82$, and 268 m^3 suggested k values from 0.3 to 1.0 depending on the ventilation rate. Higher ratios of q/V gave lower values of k (mixing less good with higher flow rate) although no explicit value of R was used (Ishizu, 1980). In tests on a 222-m^3 schoolroom, k was 0.58 (4.6 air changes/hr) and 0.5 (3 ach) for CO_2. However, these values were based on actual air changes/hour as measured by dissipation of SF_6 tracer, rather than on measured q/V (Kusuda, 1976). Use of the actual q would probably reduce the k value. With regard to mixing, it is useful to recognize that, for residences, mixing between rooms usually takes place within an hour or less, and pollutant concentrations rarely vary by more than a factor of

Table 6.8   Deposition Velocities for SO$_2$ on Wool Carpets, Wallpaper, and Paint[a]

		Deposition Velocity, K_{dep} (cm/min)	
	pH	Pile	Backing
Carpet			
Pink Tufted	4.1	1.9	0.4
Orange Tufted	3.5	3.2	0.8
Green Tufted	6.7	4.4	1.0
White Tufted	9.2	4.3	0.7
Mustard Tufted	4.8	1.3	1.0
Dark Green Looped (Unused)	4.3	1.2	0.7
Dark Green Looped (Used)	4.5	1.7	0.8
Wallpaper		*Surface*	
Embossed		5.8	
Vinyl		0.4	
Paint			
Gloss		2.0	
Emulsion		1.7	
Test Room (Wilson, 1968)		0.43–0.64	

[a]Walsh et al., 1977.

2 within the interior space. These observations are discussed in more detail in Chapter 11.

VALIDATION

Experimental validation of the completely mixed mass balance model has only been carried out for a limited number of pollutants and conditions. However, the model appears to fit satisfactorily as long as mixing is reasonably complete and the major contributions to the mass balance are sufficiently characterized. Figure 6.5 shows the decay of a tracer, SF$_6$, in a 41-m^3 room with four large fans and a single exhaust vent. The mixing factor k for these data was 0.9 (Drivas et al., 1972).

The effect of nonideal mixing is demonstrated for cigarette smoke particles in Figure 6.6 (268-m^3 room, 106-m^3/min ventilation rate, no recirculated air, 50 smouldering cigarettes) and Figure 6.7 (82-m^3 room, 62-m^3/min outdoor air ventilation rate, 15-m^3/min recirculated air rate, filter efficiency assumed

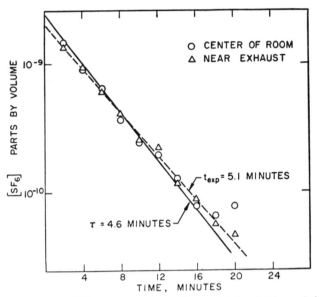

Figure 6.5. Comparison of SF$_6$ concentrations with completely mixed mass balance model for a 41-m^3 room (τ is the theoretical time for one air change) [Drivas et al. (1972). Reprinted by permission of The American Chemical Society from *Environmental Science and Technology*.]

to be zero, 6 smouldering cigarettes) (Ishizu, 1980). The particle generation rate was estimated at 1.1 mg/min per cigarette. Appropriate mixing factors appear to be between 0.3 and 0.6.

Figures 6.8 and 6.9 show comparisons between the model for ozone for a

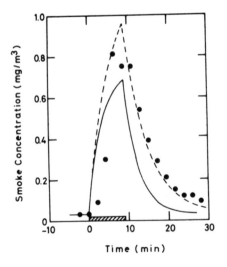

Figure 6.6. Comparison of measured and predicted cigarette smoke particle concentrations for a 268-m^3 room, with completely mixed mass balance model [Ishizu (1980). Reprinted by permission of The American Chemical Society from *Environmental Science and Technology*]: $\square\square\square$, smouldering period; •, measured values. Predicted concentrations: ——, $k = 1.0$; ---, $k = 0.6$.

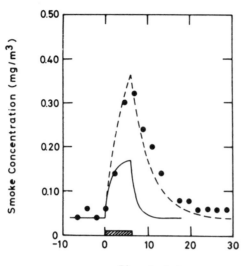

Figure 6.7. Comparison of measured and predicted cigarette smoke particle concentrations for an 82-m³ room, with completely mixed mass balance model [Ishizu (1980). Reprinted by permission of The American Chemical Society from *Environmental Science and Technology*.] : [///], smouldering period; •, measured values. Predicted concentrations: ———, $k = 1.0$; – – –, $k = 0.3$.

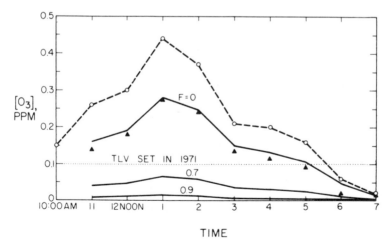

Figure 6.8. Comparison of measured and predicted indoor ozone concentrations with completely mixed mass balance model [Shair and Heitner (1974). Reprinted by permission of The American Chemical Society from *Environmental Science and Technology*.] : $F \simeq 0$; – –o– –, outdoor concentrations; ▲, indoor concentration; ———, predicted indoor concentrations; August 18, 1973, Pasadena, California.

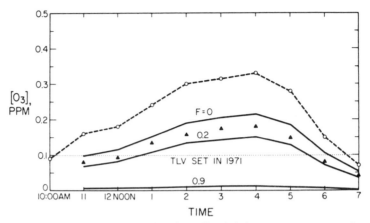

Figure 6.9. Comparison of measured and predicted indoor ozone concentrations with completely mixed mass balance model [Shair and Heitner (1974). Reprinted by permission of The American Chemical Society from *Environmental Science and Technology*.]: $F \simeq$ 0.1–0.2; --o--, outdoor concentration; ▲, indoor concentration; ——, predicted indoor concentration; August 11, 1973, Pasadena, California.

well-mixed conference room and office (V = 209 m^3 ; surface area/V = 0.49 m^{-1} ; q_0 = 19 m^3/min; q_1 = 22 m^3/min; q_2 = 0; S = 0; K_{dep} = 3 cm/min) (Shair and Heitner, 1974). Equation (6.2) was solved with a concentration-dependent term for R [i.e., Equation (6.4)]. The measured outdoor concentration was approximated with a ramp function:

$$C_o = C_{b,o} + \alpha \Delta t \tag{6.14}$$

where $C_{b,o}$ is the outdoor concentration at the beginning of a time interval, Δt, and α is an empirical constant determined from the data over Δt. Values of C_o from Equation (6.14) were input into the integrated form of Equation (6.2) over Δt. Since make-up and recirculation air passed through the same glass fiber filter, which is inefficient for O$_3$, F_1 = $F_0 \simeq$ 0. These results are shown in Figure 6.8. Figure 6.9 reflects the effect of filter efficiency as the glass fiber was replaced with a filter with $F \simeq$ 0.1 – 0.2 (Shair and Heitner, 1974).

Similar results were obtained for carbon monoxide and ozone in several hospital intensive care areas (Allen and Wadden, 1982). Figure 6.10 reflects the tendency of the one-compartment model to follow the outdoor concentration profile. Smoking, the only identified source of CO, was estimated at 2 cigarettes/person · hr and 150 mg CO/cigarette, for average 8 a.m.–5 p.m. inhalation of one person, q_0 = 30.8 m^3/min, q_1 = 5.6 m^3/min, q_2 = 0, k = 1, R = 0, F_0 = F_1 = 0, and V = 74 m^3 (Allen, 1977). The influence of the source term, although possibly underestimated, is evident in the model predictions for the daytime work period. Annual measured mean values agreed well with predicted values,

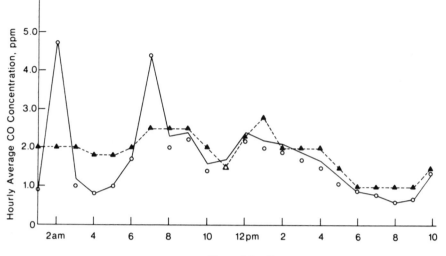

Figure 6.10. Comparison of measured and predicted indoor CO concentrations with completely mixed mass balance model (Allen and Wadden, 1982; data from Allen, 1977): o, outdoor concentration; ---▲---, indoor concentration; ——, predicted indoor concentration; July 27, 1976, Chicago, Illinois.

within 0.2 ppm for CO and 0.002 ppm for O_3. However, shorter-term comparisons were less successful. The correlation coefficient, r, between predicted and measured hourly values examined over 2-week periods, varied from 0.19 to 0.71 for CO and 0.19 to 0.91 for O_3.

MULTI-COMPARTMENT MODELS

More complex mass balance models involving the coupling of two or more spaces may be appropriate for certain specific applications. Several models of this type have been proposed (Rodgers, 1980; Scheff and Wadden, 1981; Reible and Shair, 1981) and one of these is presented here and applied in Chapter 10.

Figure 6.11 is a sketch of a two-compartment model. The nomenclature is as before save that specific values of volume, mixing factor, emission and removal rates, and concentration exist for each compartment. Implicit in the model is the assumption that air is supplied to compartment 1 only by transfer through ventilation ducts or from hallways through doors and other openings. There is no significant air exchange with the outside environment. The analysis of concentration variation in a high-use room within a large office building is one possible application of such an approach.

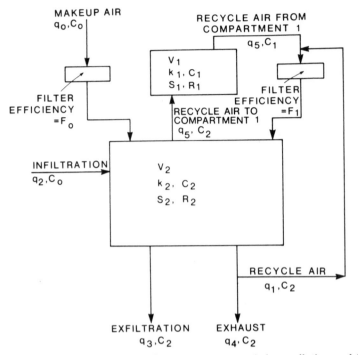

Figure 6.11.   Ventilation system for two-compartment indoor pollution model.

Referring to Figure 6.11, the mass balance equations for each of the compartments are

$$\frac{dC_1}{dt} + \alpha_3 C_1 = \alpha_2 C_2 + \alpha_1 \tag{6.15}$$

$$\frac{dC_2}{dt} + \beta_3 C_2 = \beta_2 C_1 + \beta_1 \tag{6.16}$$

where

$$\alpha_1 = \frac{S_1}{V_1}$$

$$\alpha_2 = \frac{k_1 q_5}{V_1}$$

$$\alpha_3 = \frac{(k_1 q_5 + K_1 V_1)}{V_1}$$

$$\beta_1 = \frac{S_2 + k_2 q_2 C_o + k_2(1 - F_0)q_0 C_o}{V_2}$$

$$\beta_2 = \frac{k_2 q_5(1 - F_1)}{V_2}$$

$$\beta_3 = \frac{k_2(q_1 F_1 + K_2 V_2 + q_0 + q_2)}{V_2}$$

Table 6.9   Coefficients, Constants, and Exponents for Two-Compartment Model: Equations (6.17) and (6.18)

$\alpha_1 = S_1/V_1$

$\alpha_2 = k_1 q_5/V_1$

$\alpha_3 = (k_1 q_5 + K_1 V_1)/V_1$

$\beta_1 = [S_2 + k_2 q_2 C_o + k_2(1 - F_0)q_0 C_o]/V_2$

$\beta_2 = k_2 q_5(1 - F_1)/V_2$,

$\beta_3 = k_2(q_1 F_1 + K_2 V_2 + q_0 + q_2)/V_2$

$\zeta_1 = \frac{1}{2}[(\beta_3 + \alpha_3) + \sqrt{(\beta_3 + \alpha_3)^2 - 4(\alpha_3\beta_3 - \alpha_2\beta_2)}]$

$\zeta_2 = \frac{1}{2}[(\beta_3 + \alpha_3) - \sqrt{(\beta_3 + \alpha_3)^2 - 4(\alpha_3\beta_3 - \alpha_2\beta_2)}]$

C_{1s} = steady-state solution for compartment 1 ($t \to \infty$)

$\quad = (\alpha_2/\alpha_3) [(\beta_2\alpha_1/\alpha_3 + \beta_1)/(\beta_3 - \beta_2\alpha_2/\alpha_3)] + \alpha_1/\alpha_3$

C_{2s} = steady-state solution for compartment 2 ($t \to \infty$)

$\quad = (\beta_2\alpha_1/\alpha_3 + \beta_1)/(\beta_3 - \beta_2\alpha_2/\alpha_3)$

C_{10} = concentration in compartment 1 at $t = 0$

C_{20} = concentration in compartment 2 at $t = 0$

$M_{12} = [\alpha_2 C_{20} + \alpha_1 - \alpha_3 C_{10} + \zeta_1(C_{10} - C_{1s})]/(\zeta_1 - \zeta_2)$

$M_{11} = C_{10} - C_{1s} - M_{12}$

$M_{22} = [\beta_2 C_{10} + \beta_1 - \beta_3 C_{20} + \zeta_1(C_{20} - C_{2s})]/(\zeta_1 - \zeta_2)$

$M_{21} = C_{20} - C_{2s} - M_{22}$

Note: Occasionally ζ_1 and ζ_2 will have imaginary parts. The following relationships will be helpful in developing the final form of Equations (6.17) and (6.18):

$e^{i\theta} = \cos\theta + i\sin\theta$

$\cos\theta = (e^{i\theta} + e^{-i\theta})/2$

$\sin\theta = (e^{i\theta} - e^{-i\theta})/2i$

$\cos i\theta = \cosh\theta = (e^{\theta} + e^{-\theta})/2$

$\sin i\theta = i\sinh\theta = i(e^{\theta} - e^{-\theta})/2$

The method of characteristics (e.g., Hildebrand, 1949) is used to solve the coupled first-order system of Equations (6.15) and (6.16). The solution is

$$C_1 = M_{11} e^{-\zeta_1 t} + M_{12} e^{-\zeta_2 t} + C_{1s} \tag{6.17}$$

$$C_2 = M_{21} e^{-\zeta_1 t} + M_{22} e^{-\zeta_2 t} + C_{2s} \tag{6.18}$$

where the values of coefficients, constants, and exponents are given in Table 6.9 (Scheff and Wadden, 1981).

The usefulness of multi-compartment models depends on whether these more complex predictive techniques are able to provide greater discrimination of indoor pollutant distributions. In a recent study of a laboratory–office building, SF_6 was used as a tracer to follow the concentration history at 11 separate locations (Reible and Shair, 1981). Two physically separated rooms (based on blueprints of the building's ventilation system), which apparently could only interact through a small air handler, were modeled using one-, two-, and three-compartment models. During the course of the study it was discovered that while the forced-ventilation system for the building was in balance, there was a significant design flow imbalance for the rooms being analyzed. Comparison of the modeling results with measured concentrations suggested that the one-compartment model appeared to be as satisfactory as either of the multi-compartment forms, although all the prediction schemes gave results within the scatter of the data.

EMPIRICAL MODELS

These types of models are based on statistical evaluation of concurrent indoor and outdoor measurements of concentrations along with other variables which may be considered to be of importance. Ordinarily the data are analyzed using partial correlation and linear regression procedures. The simplest form would then be

$$C_i = \beta_1 C_o + \beta_0 \tag{6.19}$$

where β_1 and β_0 are empirically determined from least-squares fitting of the data (e.g., Kleinbaum and Kupper, 1978). The values of C_o and C_i are those averaged over the measurement times or may be longer-term averages. An example of an empirical model is Equation (4.4) for formaldehyde.

When appropriate data are available, it is sometimes possible to fit a deterministic model with empirically derived constants, particularly when many of the factors in Equation (6.3) are not available. An example of this approach was described by Dockery and Spengler (1981b) for indoor concentrations of respirable particles and respirable sulfates. Their form of Equation (6.3) is

$$\overline{C}_i = (1 - F)\overline{C}_o + \frac{\overline{S}}{q_0 + q_2} = P\overline{C}_o + \frac{\overline{S}}{q} \tag{6.20}$$

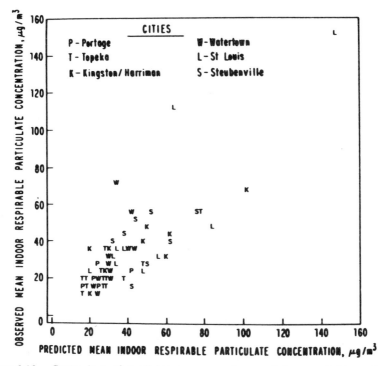

Figure 6.12. Comparison of predicted vs. measured mean indoor respirable particulate concentrations for each home. Twelve observations are hidden (Dockery and Spengler, 1981b).

where the overbar indicates 24-hr average values. The estimated indoor concentration, $\overline{C}_i$, is then determined by the measured outdoor concentration and two empirically determined constants: P, the penetration of outdoor pollutants indoors, and $\overline{S}/q$, the indoor source term. Variations in P and $\overline{S}/q$ were not considered for estimation of annual mean indoor concentrations. It was assumed that P was a simple linear function of air conditioning:

$$P = \beta_1 + \beta_2(A) \tag{6.21}$$

where A was an indicator variable for fully air conditioned sites and the β's were empirically determined. For particles,

$$\frac{\overline{S}}{q} = \beta_3(N_{cig}) + \beta_4(AN_{cig}) + \beta_5(A) + \beta_0 \tag{6.22}$$

where N_{cig} was the estimated number of cigarettes smoked indoors per day. For sulfates,

$$\frac{\overline{S}}{q} = \beta_3(N_{cig}) + \beta_4(AN_{cig}) + \beta_5(G) + \beta_6(AG) + \beta_7(A) + \beta_0 \qquad (6.23)$$

where G is an indicator variable for cooking fuel (1 for gas, 0 otherwise) and the β's were determined by regression on indoor–outdoor concentration data for 68 sites. The indoor respirable particle model had an overall correlation of 0.68 and the comparison between observed and predicted mean indoor concentration values is shown in Figure 6.12. The sulfate model had an overall multiple correlation of 0.79 and the predicted–observed pattern is shown in Figure 6.13. The calculated β values are given in Tables 6.10 and 6.11.

Another application of empirical–deterministic modeling has been to estimate personal exposures. Using a subset of the sulfate and respirable particle data described above, along with information on time spent indoors and outdoors, Dockery and Spengler (1981a) postulated a time-weighted model for particles,

$$E = \frac{t_{home}C_i + t_{out}C_o + t_{other}(C_i + N_{smoke} \times 20\,\mu g/m^3)}{t_{tot}} \qquad (6.24)$$

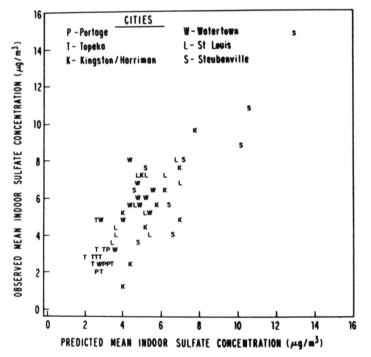

Figure 6.13. Comparison of predicted vs. measured mean indoor respirable sulfate concentrations for each home. Twelve observations are hidden (Dockery and Spengler, 1981b).

Table 6.10   Respirable Particles Regression Analysis Results[a]

Parameter	Variable	Regression Coefficient	Standard Error of Coefficient	F Value
β_1	$\overline{C}_o$	0.70	0.035	387.55[b]
β_2	$A\overline{C}_o$	−0.39	0.047	69.17[b]
β_3	N_{cig}	0.88	0.057	242.96[b]
β_4	AN_{cig}	1.23	0.079	243.79[b]
β_5	A	−2.39	1.835	1.70
β_0	Constant	15.02	0.882	—

[a]Dockery and Spengler (1981b).
[b]Probability of $F < 0.001$.

and for sulfates,

$$E = \frac{t_{\text{in}} C_i + t_{\text{out}} C_o}{t_{\text{tot}}} \tag{6.25}$$

where E is the estimated exposure in $\mu g/m^3$; t_{home}, time spent at home; t_{out}, time outdoors; t_{in}, time indoors; t_{other}, time spent indoors away from home; t_{tot}, total sample length; N_{smoke}, number of smokers a participant was exposed to away from home; C_i average indoor home concentration; and C_o average outside concentration in $\mu g/m^3$. Comparisons for 37 volunteers between predicted and measured values gave reasonable estimates; but these were only slightly better than statistical estimates using measured indoor concentrations.

Table 6.11   Respirable Sulfate Regression Analysis Results[a]

Parameter	Variable	Regression Coefficient	Standard Error of Coefficient	F Value
β_1	$\overline{C}_o$	0.75	0.012	3839.69[b]
β_2	$A\overline{C}_o$	−0.47	0.019	613.81[b]
β_3	N_{cig}	0.006	0.007	0.82
β_4	AN_{cig}	0.046	0.011	16.52[b]
β_5	G	1.08	0.142	58.62[b]
β_6	AG	−0.97	0.282	11.78[b]
β_7	A	1.73	0.212	66.63[b]
β_0	Constant	0.04	0.092	—

[a]Dockery and Spengler (1981b).
[b]Probability of $F < 0.001$.

REFERENCES

Achenbach, P. R. and Coblentz, C. W. (1963). Field measurements of air infiltration in ten electrically-heated houses. *ASHRAE Trans.* **69**:358–365.

Allen, R. J. (1977). *Relationship between indoor and outdoor concentrations of carbon monoxide and ozone for an urban hospital.* Ph.D. Thesis, University of Illinois at the Medical Center, Chicago.

Allen, R. J. and Wadden, R. A. (1982). Analysis of indoor concentrations of carbon monoxide and ozone for an urban hospital. *Environ. Res.* **27**:136–149.

Allen, R. J., Wadden, R. A., and Ross, E. D. (1978). Characterization of potential indoor sources of ozone. *Am. Ind. Hyg. Assn. J.* **39**:466–471.

ASHRAE (1981). *ASHRAE Handbook: 1981 Fundamentals,* American Society of Heating, Refrigerating and Air Conditioning Engineers, New York.

Blomsterberg, A. K. and Harrje, D. T. (1979). Approaches to evaluation of air infiltration energy losses in buildings. *ASHRAE Trans.* **85**(Part 1):797–815.

Caffey, G. E. (1979). Residential air infiltration. *ASHRAE Trans.* **85**(Part 1): 41–57.

Cole, J. T. Zawacki, T. S., Elkins, R. H., Zimmer, J. W., and Macriss, R. A. (1980). Application of a generalized model of air infiltration to existing homes. *ASHRAE Trans.* **86**(Part 2):765–771.

Cote, W. A. and Holcombe, J. K. (1971). The influence of air conditioning systems on indoor pollutant levels. Paper 7, Session V. First Conference on Natural Gas Research and Technology. Institute of Gas Technology, Chicago.

Drivas, P. J., Simmonds, P. G., and Shair, F. H. (1972). Experimental characterization of ventilation systems in buildings. *Environ. Sci. Technol.* **6**:609–614.

Dockery, D. W. and Spengler, J. D. (1981a). Personal exposure to respirable particulates and sulfates. *J. Air Pollut. Control Assoc.* **31**:153–159.

Dockery, D. W. and Spengler, J. D. (1981b). Indoor–outdoor relationships of respirable sulfates and particles. *Atmos. Environ.* **15**:335–343.

Grimsrud, D. T., Sherman, M. H., Blomsterberg, A. K., and Rosenfeld, A. H. (1979). *Infiltration and air leakage comparisons: conventional and energy efficient housing designs.* Lawrence Berkeley Laboratory Report No. LBL-9157, University of California.

Grimsrud, D. T., Sonderegger, R. C., Sherman, M. H., Diamond, R. C., and Blomsterberg, A. (1980). *Calculating Infiltration: Implications for a Construction Quality Standard.* Lawrence Berkeley Laboratory Report No. LBL-9416, University of California.

Grot, R. A. and Clark, R. E. (1979). Air leakage characteristics and weatherization techniques for low-income housing. Presented at DOE/ASHRAE Conference on Thermal Performance of Exterior Envelopes of Buildings, Orlando, Florida, December. (See also NAS, 1981.)

Hildebrand, F. B. (1949). *Advanced Calculus for Engineers*, Prentice-Hall, Englewood Cliffs, N.J.

Hoegg, U. R. (1972). Cigarette smoke in closed spaces. *Environ. Health Perspec.* 2:117–128.

Ishizu, Y. (1980). General equation for the estimation of indoor pollution *Environ. Sci. Technol.* 14:1254–1257.

Janssen, J. E., Pearman, A. N., and Hill, T. J. (1980). Calculating infiltration: An examination of handbook models. *ASHRAE Trans.* 86(Part 2):751–764.

Kleinbaum, D. G. and Kupper, L. L. (1978). *Applied Regression Analysis and Other Multivariate Methods*, Duxbury Press, North Scituate, Mass.

Kusuda, T. (1976). Control of ventilation to conserve energy while maintaining acceptable indoor air quality. *ASHRAE Trans.* 82(Part I):1169–1181.

Malik, N. (1978). Field studies of dependence of air infiltration on outside temperature and wind. *Energy and Buildings* 1:281–292.

Mattingly, G. E. and Peters, E. F. (1977). Wind and trees: Air infiltration effects on energy in housing. *J. Ind. Aerodyn.* 2:1–19.

Moschandreas, D. J., Stark, J. W. C., McFadden, J. C., and Morse, S. S. (1978). *Indoor Air Pollution in the Residential Environment*, Vols. I and II. U.S. Environmental Protection Agency Report No. EPA 600/7-78-229a and b, Research Triangle Park, N.C.

Mueller, F. X., Loeb, L., and Mapes, W. H. (1973). Decomposition rates of ozone in living areas. *Environ. Sci. Technol.* 7:342–346.

NAS (1981). *Indoor Air Pollutants*. National Academy of Sciences, Washington, D.C.

Reible, D. D. and Shair, F. H. (1981). The reentrainment of exhausted pollutants into a building due to ventilation system imbalance. Division of Chemistry and Chemical Engineering, California Institute of Technology, Pasadena, California. International Symposium on Indoor Air Pollution, Health and Energy Conservation, University of Massachusetts, Amherst, Massachusetts, October.

Rodgers, L. C. (1980). Air quality levels in a two-zone space. *ASHRAE Trans.* 86(Part 2):92–98.

Roseme, G. D., Hollowell, C. D., Meier, A., Rosenfeld, A., and Turiel, I. (1979). Air-to-air heat exchangers: Saving energy and improving indoor air quality. In: *Changing Energy Use Futures*, R. A. Fazzolare and C. B. Smith, Eds. Vol. 3, Pergamon, New York.

Sabersky, R. H., Sinema, D. A., and Shair, F. H. (1973). Concentrations, decay rates and removal of ozone and their relation to establishing clean indoor air. *Environ. Sci. Technol.* 7:347–353.

Scheff, P. A. and Wadden, R. A. (1981). Source emissions and predictive modeling of indoor air pollution. University of Illinois School of Public Health, Chicago, Illinois. International Symposium on Indoor Air Pollution, Health and Energy Conservation, University of Massachusetts, Amherst, Massachusetts, October.

Selway, M. D., Allen, R. J., and Wadden, R. A. (1980). Ozone production from photocopying machines. *Am. Ind. Hyg. Assn. J.* 41:455–459.

Shair, F. H. (1981). Relating indoor pollutant concentrations of ozone and sulfur dioxide to those outside: Economic reduction of indoor ozone through selective filtration of air. *ASHRAE Trans.* 87(Part I):116–139.

Shair, F. H. and Heitner, K. L. (1974). Theoretical model for relating indoor pollutant concentrations to those outside. *Environ. Sci. Technol.* 8:444–451.

Shaw, C. Y. (1980). Wind and temperature induced pressure differentials and an equivalent pressure difference model for predicting air infiltration in schools. *ASHRAE Trans.* 86(Part 1):268–279.

Shaw, C. Y. (1981). Air tightness. Supermarkets and shopping malls. *ASHRAE J.*, pp. 44–46, March.

Sherman, M. H. And Grimsrud, D. T. (1980a). *Measurement of Infiltration Using Fan Pressurization and Weather Data.* Lawrence Berkeley Laboratory Report No. LBL-10852, University of California.

Sherman, M. H. and Grimsrud, D. T. (1980b). Infiltration–Pressurization correlation: Simplified physical modeling. *ASHRAE Trans.* 86(Part 2):778–807.

Sutton, D. J. Nodolf, K. M., and Makino, K. K. (1976). Predicting ozone concentrations in residential structures. *ASHRAE J.*, pp. 21–26, September.

Tamura, G. T. (1975). Measurement of air leakage characteristics of house enclosures. *ASHRAE Trans.* 81(Part 1):202–211.

Turk, A. (1963). Measurements of odorous vapors in test chambers: Theoretical. *ASHRAE J.*, pp. 55–58, October.

Wade, W. A., Cote, W. A., and Yocum, J. E. (1975). A study of indoor air pollution. *J. Air Pollut. Control Assoc.* 25:933–939.

Walsh, M., Black, A., and Morgan, A. (1977). Sorption of SO_2 by typical indoor surfaces including wool carpets, wallpaper, and paint. *Atmos. Environ.* 11:1107–1111.

Wilson, M. J. G. (1968). Indoor air pollution. *Proc. Soc. Lond. Ser. A.* 300:215–221.

Woods, J. E., Maldonado, E. A. B., and Reynolds, G. L. (1981). Safe and energy efficient control strategies for indoor air quality. Energy Research Institute, Iowa State University, Ames, Iowa, BEUL 81-01. Meeting of the American Association for the Advancement of Sciences, Toronto, January.

PART THREE

INDOOR AIR
QUALITY CONTROL

7

INDOOR AIR
QUALITY CONTROL

INDOOR AIR QUALITY STANDARDS

Most indoor air quality standards for toxic pollutants are designed for occupational settings. These standards are based on protecting healthy workers exposed to time-weighted average concentrations less than, or equal to, specified levels for up to 8 hr/day, 40 hr/week. A sampling of recommended occupational standards are shown in Table 1.3. These standards are not satisfactory for general use because they do not protect susceptible populations and are limited to 40 hr/week. Acceptable air quality for nonoccupational settings is generally based on outdoor ambient air standards or specified indoor ventilation rates. Federal air quality standards shown in Table 1.2 are designed to protect susceptible populations from adverse effects from exposure to air pollution. ASHRAE recommends that these standards also be applied to nonoccupational indoor settings. Table 7.1 is a list of ambient air quality standards for substances not covered in Table 1.2 (ASHRAE, 1980). Although the rationales for some of these standards are less well defined than those for ambient air, they are intended to protect susceptible populations from adverse health effects and are, therefore, appropriate as indoor standards.

ASHRAE standard 62-73R is a proposed standard for minimum ventilation required to achieve acceptable indoor air quality. It is a proposed revision of ASHRAE standards 62-73 (ASHRAE, 1973) and 90-75 (ASHRAE, 1975), Standard 62-73 specified minimum and recommended ventilation rate values. Because of the need for energy efficient building, in 1975 ASHRAE amended 62-73 with standard 90-75 "Energy Conservation in New Building Design." This standard, which has been incorporated into building codes of 45 states (Woods et al., 1981), specifies that the minimum values, and not the recommended values, of 62-73 be used for design purposes. ASHRAE 62-73R has been designed to resolve the differences between these two standards. Most current building codes, however, are still based on ASHRAE 90-75.

Table 7.1 Additional Ambient Air Quality Standards[a]

Contaminant[b]	Long Term Level[c]	Long Term Time	Short Term Level	Short Term Time
Acetone–O	7 mg/m³	24 hr	24 mg/m³	30 min
Acrolein–O			25 µg/m³	C[d]
Ammonia–O	0.5 mg/m³	yr	7 mg/m³	C
Beryllium	0.01 µg/m³	30 days		
Cadmium	2.0 µg/m³	24 hr		
Calcium oxide (lime)			20–30 µg/m³	C
Carbon disulfide–O	0.15 mg/m³	24 hr	0.45 mg/m³	30 min
Chlorine–O	0.1 mg/m³	24 hr	0.3 mg/m³	30 min
Chromium	1.5 µg/m³	24 hr		
Cresol–O	0.1 mg/m³	24 hr		
Dichloroethane–O	2.0 mg/m³	24 hr	6.0 mg/m³	30 min
Ethyl acetate–O	14 mg/m³	24 hr	42 mg/m³	30 min
Formaldehyde–O			150 µg/m³	C
Hydrochloric acid–O	0.4 mg/m³		3 mg/m³	30 min
Hydrogen sulfide–O	40–50 µg/m³	24 hr	42 µg/m³	1 hr
Mercaptans–O			20 µg/m³	1 hr
Mercury	2 µg/m³	24 hr		
Methyl alcohol (methanol)–O	1.5 mg/m³	24 hr	4.5 mg/m³	30 min
Methylene chloride–O	20 mg/m³	yr	150 mg/m³	30 min
	50 mg/m³	24 hr		
Nickel	2 µg/m³	24 hr		
Nitric oxide	0.5 mg/m³	24 hr	1 mg/m³	30 min
Phenol–O	0.1 mg/m³	24 hr		
Sulfates	4 µg/m³	yr		
	12 µg/m³	24 hr		
Sulfuric acid–O	50 µg/m³	yr	200 µg/m³	30 min
	100 µg/m³	24 hr		
Trichlorethylene–O	2 mg/m³	yr	16 mg/m³	30 min
	5 mg/m³	24 hr		
Vanadium	2 µg/m³	24 hr		
Zinc	50 µg/m³	yr		
	100 µg/m³	24 hr		

[a]ASHRAE (1980).
[b]The materials marked "O" have odors at concentrations sometimes found in outdoor air. The tabulated concentration levels do not necessarily result in odorless conditions. (See Table 7.4.)
[c]Unless otherwise specified, all air quality measurements should be corrected to standard conditions of 25°C (77°F) temperature and 760 mm (29.92 in.) of mercury pressure.
[d]Ceiling, or maximum allowable concentration.

Standard 62-73R specifies both indoor air quality levels and minimum ventilation rates which will be acceptable to human occupants and will not impair health. The recommendations apply to all occupied indoor spaces not covered by occupational standards. Acceptable air quality is defined as air with no known contaminants at harmful concentrations and with which a substantial majority of the people exposed do not express dissatisfaction. Two procedures to obtain acceptable air quality indoors are specified. Acceptable indoor air quality can be achieved by providing ventilation air of proper quality and quantity to the space; or by not exceeding concentration standards for identified contaminants.

The ventilation procedure of 62-73R includes ventilation rate and air clean-up criteria. The standard is based on physiological considerations, subjective evaluation, and professional judgment. When applying the standard, the use of 100% outdoor air of acceptable quality for make-up air is assumed. However, outdoor air quantities can be reduced when recirculated air is properly treated by air clean-up devices. Outdoor air is considered nonacceptable for ventilation if it contains any contaminant at concentrations above those in Tables 1.2 and 7.1. If outdoor air is known to contain a contaminant not listed in these tables, the concentration should not exceed $\frac{1}{10}$ of the occupational standard. If outdoor air is known to exceed any of these standards, it must be treated before it is used to control the offending contaminants. Indoor air quality will be acceptable if ventilation rates shown in Table 7.2 are provided. Note that higher rates are specified for spaces where smoking is permitted.

A minimum supply of air is required to dilute the CO_2 produced by metabolism. It is generally believed that 0.5% CO_2 is a reasonable limit at which people can function without any adverse effects (see section on CO_2 in Chapter 2). Assuming an outdoor concentration of 0.03% and an activity level of 1 met (50 kcal/m^2 · hr), 2.25 cfm/person is required to keep indoor CO_2 below 0.5%. One met is a representative activity level of a sedentary adult office worker. ASHRAE recommends an additional safety factor of 2 and that the *minimum outdoor air requirement be 5 cfm/person*. This level is linear with metabolic activity and correspondingly higher limits are needed where activity is greater than 1 met (see Tables 4.16 and 7.3). If exhaust air is used as supply air, it must be equivalent to acceptable outdoor air, but the outdoor air portion must never be less than 5 cfm/person. The suggested standard also specifies criteria by which ventilation systems can be shut off for spaces not continuously occupied.

In addition to the toxicological and physiological basis of the ASHRAE standard, indoor ventilation systems are sometimes designed on the basis of heat and odor control criteria. Heat control standards are based on heat balance equations and include terms for metabolism, evaporation, radiation, and convection. These standards and procedures have been developed for, and are ordinarily only applied to, industrial settings. Figure 7.1 shows the relationships between metabolism, evaporation, radiation and convection, and storage; and heat loss or gain

Table 7.2 Outdoor Air Requirements for Ventilation[a]

	Estimated Occupancy, Persons per 1000 ft² or 100 m² Floor Area. Use Only When Design Occupancy Is Not Known	Outdoor Air Requirements				Comments
		Smoking	Nonsmoking	Smoking	Nonsmoking	
		cfm/person		(L/s)/person		
Commercial Facilities						
Dry Cleaners and Laundries						
Commercial	10	–	15	–	7.5	A blank (–) indicates that smoking (or nonsmoking) in a space should not occur
Storage/pick-up areas	30	35	10	17.5	5	
Coin-operated laundries	20	35	15	17.5	7.5	Dry cleaning processes may require more air
Coin-operated dry cleaning	20	–	15	–	7.5	
Food and Beverage Services						
Dining rooms	70	35	7	17.5	3.5	
Kitchens	20	–	10	–	5	
Cafeterias, fast food facilities	100	35	7	17.5	3.5	
Bars and cocktail lounges	100	50	10	25	5	
		cfm/ft² floor		(L/s)/m² floor		
Garages, Auto Repair Shops, Service Stations						
Parking garages (enclosed)	–	1.5	1.5	0.07	0.07	Distribution must consider worker location and concentration of running engines; stands where engines are run must incorporate system for positive engine exhaust withdrawal
Auto repair workrooms (general)	–	1.5	1.5	0.07	0.07	

Hotels, Motels, Resorts, Dormitories, and Correctional Facilities

		cfm/room		(L/s)/room		Comments
						See also food and beverage services, merchandising, barber and beauty shops, garages
Bedrooms (single, double)	5	30	15	15	7.5 }	
Living rooms (suites)	20	20	10	10	5	Independent of room size
Baths, toilets (attached to bedrooms)		50	50	25	25	Independent of room size; installed capacity for intermittent use

		cfm/person		(L/s)/person		
Lobbies	30	15	5	7.5	2.5	
Conference rooms (small)	50	35	7	17.5	3.5	
Assembly rooms (large)	120	35	7	17.5	3.5	
Gambling casinos	120	35	7	17.5	3.5	

Offices

		cfm/person		(L/s)/person		
Office space	7	20	5	10	2.5	
Meeting and waiting spaces	60	35	7	17.5	3.5	

Public Spaces

		cfm/ft² floor		(L/s)/m² floor		
Corridors and utility rooms		0.02	0.02	0.001	0.001	

		cfm/stall or urinal		(L/s)/stall or urinal		
Public restrooms	100	75	—	37.5	—	

		cfm/locker		(L/s)/locker		
Locker and dressing rooms	50	35	15	17.5	7.5	

Table 7.2 *(Continued)*

	Estimated Occupancy, Persons per 1000 ft² or 100 m² Floor Area. Use Only When Design Occupancy Is Not Known	Outdoor Air Requirements				Comments
		Smoking	Nonsmoking	Smoking	Nonsmoking	
		cfm/person		(L/s)/person		
Commercial Facilities						
Retail Stores						
Sales floors and showrooms						
Basement and street floors	30	25	5	12.5	2.5	
Upper floors	20	25	5	12.5	2.5	
Storage areas (serving sales and storerooms)	15	25	5	12.5	2.5	
Dressing rooms	–	25	5	12.5	3.5	
Malls and arcades	20	10	5	5	2.5	
Shipping and receiving areas	10	10	5	5	2.5	
Warehouses	5	10	5	5	2.5	
Elevators	–	–	15	–	7.5	
Smoking rooms	70	50	–	25	–	
Speciality Shops		cfm/person		(L/s)/person		
Barber and beauty shops	25	35	20	17.5	10	
Reducing salons, health spas (exercise rooms)	20	–	15	–	7.5	
Florists	10	25	5	12.5	2.5	
Greenhouses	1	–	5	–	2.5	
Shoe repair shops (combined workrooms/trade areas)	10	15	10	7.5	5	

Application	cfm/ft² floor	cfm/person	(L/s)/m² floor	(L/s)/person	Comments
Pet shops	1		0.05		
Sports and Amusement Facilities					
Ballrooms and discos	100	35 / 7		17.5 / 3.5	
Bowling alleys (seating area)	70	35 / 7		17.5 / 3.5	
Playing floors (gymnasium)	30	– / 20		– / 10	When internal combustion engines are operated, increased ventilation rates will be required
Spectator areas	150	35 / 7		17.5 / 3.5	
Game rooms (e.g., card and billiard rooms)	70	35 / 7		17.5 / 3.5	
Swimming pools	*(cfm/ft² area)*		*((L/s)/m² area)*		
Pool and deck areas	0.5		0.025		
Spectator area	70	35 / 7		17.5 / 3.5	
Theaters					
Ticket booths		20 / 5		10 / 2.5	
Lobbies, foyers, and lounges, auditoriums in motion picture theaters, lecture, concert and opera halls	150	35 / 7		17.5 / 3.5	
Stages, TV and movie studios	70	– / 10		– / 5	Special ventilation will be needed to eliminate special stage effects (e.g., dry ice vapors, mist, etc.)

Table 7.2 (*Continued*)

	Estimated Occupancy, Persons per 1000 ft² or 100 m²/Floor Area. Use Only When Design Occupancy Is Not Known	Outdoor Air Requirements				Comments
		cfm/person		(L/s)/person		
		Smoking	Nonsmoking	Smoking	Nonsmoking	
Transportation		*Commercial Facilities*				
Waiting rooms, ticket and baggage areas, corridors and gate areas, platforms, concourses	150	35	7	17.5	3.5	Ventilation within vehicles will require special consideration
Workrooms		cfm/person		(L/s)/person		
Meat processing rooms	10	—	5	—	2.5	Spaces maintained at low temperatures (−10°F to +50°F, or −23°C to +10°C) are not covered by these requirements unless the occupancy is continuous. Ventilation from adjoining spaces is permissible. When the occupancy is intermittent, infiltration will normally exeed the ventilation requirement.
Pharmacists' workroom	20	—	7	—	3.5	
Bank vaults	10	—	5	—	2.5	
Photo studios						
Camera room, stages	10	—	5	—	2.5	
Darkrooms	10	—	20	—	10	

Application	Estimated Max Occupancy, P/1000 ft²	cfm/person (or /bed)	cfm/ft² floor	(L/s)/person (or /bed)	(L/s)/m² floor	Comments
Duplicating and printing rooms	—	—	0.5	—	0.025	Installed equipment may incorporate positive exhaust and control (as required) of undesirable contaminants (toxic or otherwise)
Institutional Facilities[b]						
Educational Facilities						
Classrooms	50	25	5	12.5	2.5	Special contaminant control systems may be required for processes or functions including laboratory animal occupancy
Laboratories	30	—	10	—	5	
Training shops	30	35	7	17.5	3.5	
Music rooms	50	35	7	17.5	3.5	Special requirements or codes and pressure relationships may determine minimum ventilation rates and filter efficiency
Libraries	20	—	5	—	2.5	
Hospitals, Nursing and Convalescent Homes						
Patient rooms	10	35 (/bed)	7	17.5 (/bed)	3.5	
Medical procedure areas	10	35	7	17.5	3.5	Activities generating contaminants may require higher rates
Operating rooms, delivery rooms	20	—	10	—	10	
Recovery and intensive care rooms	30	—	15	—	7.5	
Autopsy rooms	20	—	60	—	30	Air shall not be recirculated
Physical therapy areas	20	—	15	—	7.5	

Table 7.2 (Continued)

	Estimated Occupancy, Persons per 1000 ft^2 or 100 m^2/Floor Area. Use Only When Design Occupancy Is Not Known	Outdoor Air Requirements				Comments
		Smoking	Nonsmoking	Smoking	Nonsmoking	
Residential Facilities[c]						
		cfm/room		(L/s)/room		Operable windows or mechanical ventilation systems shall be provided for use when occupancy is greater than usual conditions or when unusual contaminant levels are generated within the space
Homes						
General living areas		10		5		
Bedrooms		10		5		
All other rooms		10		5		Independent of room size
Kitchens		100		50		Independent of room size; installed capacity for intermittent use, may be a window of 2 ft^2 (0.2 m^2) or greater opening area, with no closeable door between it and the kitchen range or bath
Baths, toilets		50		25		

	cfm/car space	(L/s)/car space	
Garages (separate for each dwelling unit)	10	5	Independent of room size
	cfm/ft² floor	(L/s)/m² floor	
Garages (common for several units)	1.5	0.07	

Industrial Facilities[d]

Industry Levels		cfm/person		(L/s)/person		
High activity level (2.5 mets)	—	35	20	17.5	10	Mining, foundry, etc.
Medium activity level (2.0 mets)	—	35	10	17.5	5	Automotive repair, assembly line, etc.
Low activity level (1.5 mets)	—	35	7	17.5	.5	Laboratory work, light assembly, etc.

[a]ASHRAE (1980).
[b]For areas not listed, refer to Commercial Facilities.
[c]Private dwelling places, single or multiple, low or high rise.
[d]Occupational safety laws usually regulate process ventilation requirements. The list gives the requirements for the occupants only, assuming that the ventilated air is of an appropriate quality.

Table 7.3   Heat Rejection for Various Activity Levels

Activity	Heat Rejection	
	Btu/hr[a]	Number of Mets[b]
Sleeping	280	0.7
Seated at rest	400	1.0
Standing relaxed	480	1.2
Shopping	560–720	1.4–1.8
Housework	640–1360	1.6–3.4
Office work	480–560	1.2–1.8
Light factory work	800–960	2.0–2.4
Heavy factory work	1400–1800	3.5–4.5
Dancing	960–1760	2.4–4.4

[a]Olivieri (1979).
[b]Based on 2 m^2/person; 1 met = 50 kcal/hr · m^2 = metabolic rate of a person sitting at rest in comfortable environment.

for a clothed subject at rest for varying dry bulb temperatures and constant relative humidity (ACGIH, 1980). In order to maintain thermal equilibrium, the body must lose and produce heat at exactly the same rate. This can range from 400 Btu/hr for sedentary adults to up to 4000 Btu/hr under heavy exertion. Heat rejection values for various activities are given in Table 7.3 and some appropriate ventilation volumes are given at the end of Table 7.2.

Odor control is incorporated into ASHRAE 62-73R through the statement

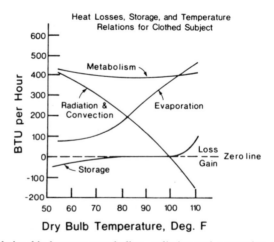

Figure 7.1.   Relationship between metabolism, radiation and convection, evaporation, and storage on heat loss [ACGIH (1980). Reprinted by permission of The Committee on Industrial Ventilation, ACGIH, Lansing, Michigan.]

that air is free of annoying contaminants if at least 80% of a panel of untrained observers deems the air to be not objectionable (ASHRAE, 1980). Characterization of odor thresholds and evaluation of the reaction of observers to odor quality are problems of considerable complexity. Table 7.4 lists the recognition thresh-

Table 7.4   Selected Odor Recognition Thresholds

| Compound | Odor Recognition Threshold (ppm) | | Odor Description |
	Leonardos et al. (1969)	Hellman and Small (1974)	
Acetaldehyde	0.21		Green sweet
Acetic acid	1.0		Sour
Acetone	100	140	Chemical sweet; pungent; fruity
Acrolein	0.21		Burnt sweet; pungent
Ammonia	46.8		Pungent
Amyl acetate, primary		0.21	Sweet; ester; banana
Amyl alcohol		1.0	Sweet
Benzene	4.68		Solvent
Butyl cellosolve acetate		0.20	Sweet; ester
Butyric acid	0.001		Sour
Carbon tetrachloride	21.4–100		Sweet; pungent
Cellosolve solvent		1.3	Sweet; musty
1-4 Dioxane		5.7	Sweet; alcohol
Ethyl acetate		13.2	Sweet; ester
Ethyl acrylate	0.00047	0.00036	Hot plastic; earthy; sour; pungent
Ethylene		700	Olefinic
Ethyl mercaptan	0.001		Earthy; sulfidy
Formaldehyde	1.0		Hay/strawlike; pungent
Hydrogen sulfide gas	0.0047–0.00047		Eggy sulfide
Isobutyraldehyde		0.236	Sweet; ester
Methanol	100	53.3	Sweet; sour; sharp
Methyl cellosolve acetate		0.64	Sweet; ester
Methylene chloride	214		
Methyl ethyl ketone	10.0	6.0	Sweet; sharp
Methyl isobutyl ketone	0.47	0.28	Sweet; sharp
Methyl mercaptan	0.0021		Sulfidy; pungent
Methyl methacrylate	0.21	0.34	Sulfidy; pungent; sweet; sharp
Perchloroethylene	4.68		Chlorinated solvent
Phenol	0.047		Medicinal
Pyridine	0.021		Burnt; pungent; diamine
Styrene (inhibited)	0.1	0.15	Solventy; rubbery; sharp; sweet
Sulfur dioxide	0.47		
Toluene (petroleum)	2.14	1.74	Mothballs; rubbery; sour; burnt
Trichloroethylene	21.4		Solventy
Trimethylamine	0.00021		Fishy; pungent
Xylene		0.27	Sweet

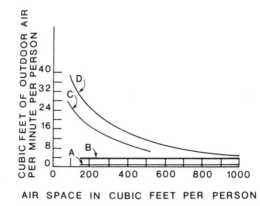

AIR SPACE IN CUBIC FEET PER PERSON

Figure 7.2. Ventilation requirements for O_2, CO_2, and odor control [HPAC (1942). Reprinted by permission of Heating, Piping and Air Conditioning.]

A. Air required to provide necessary O_2 content;
B. Air required to prevent CO_2 concentration from rising above 0.5%;
C. Air required to remove objectionable body odors from sedentary adults;
D. Curve C increased by 50% to allow for moderate physical activity.

olds for a variety of materials, based on 100% response from panels of trained observers (Leonardos et al., 1969; Hellman and Small, 1974). Of special note are those common odors, such as butyric acid and ammonia, associated with human activity (see Table 4.16 for typical bioeffluents). In general, response to odor intensity is usually considered to be, at least roughly, a function of the log of the concentration of the compound being considered.

Figure 7.2 shows ventilation requirements for supplying sufficient oxygen, for CO_2 control and for odor control. It is apparent that ventilation for odor control, for materials normally present with human inhabitation, is a function of air space per person and can be many times greater than ventilation required for CO_2 control.

VENTILATION SYSTEM DESIGN

Ventilation systems are designed to meet a wide range of criteria. The objectives of ventilation control include the supply of sufficient O_2 for normal respiration, dilution of contaminants within occupied spaces, removal of contaminants emitted from point sources or work areas (local exhaust), pressurization of spaces to control infiltration or exfiltration, dissipation of thermal loads in occupied spaces, and odor control (Woods et al., 1981). The supply of O_2 is usually not a problem if ventilation is provided for dilution, pressurization, or heat dissipation (Woods, 1980). The use of dilution ventilation for health hazard control of a contaminant should not be used if the quantity generated is very great, local concentrations can build-up, the toxicity is high, or the emission is not uniform (ACGIH, 1980). Large, nonuniform releases of toxic substances will require impractically large volumes of dilution air to prevent hazardous indoor concentrations. The use of properly designed local exhausts near major sources of toxic contaminants may be a far more cost-effective solution to the control of indoor air quality.

Some general principles of dilution ventilation are illustrated in Figure 7.3. The first six designs are examples of poor fan locations. In each of these rooms, the exhaust air duct is above and behind the work surface. Smoke, fumes or gases emitted at the workbench will be drawn into the worker's breathing zone. The bottom six diagrams all have the exhaust duct over the work area. In this location, the fan will provide local exhaust for the work bench in addition to general room air exhaust. Figure 7.3 also demonstrates correct and incorrect inlet air or make-up air locations. Poor inlet air locations either short circuit directly to the exhaust or push the contaminants from the bench into the room. Fair inlet locations do not mix workbench emissions with room supply air. Good supply air inlets have forced make-up air systems. The best air inlet uses a plenum to distribute make-up air over a large area.

The basic principles to be applied to dilution ventilation are (ACGIH, 1980):

1. Supply to a room the amount of air required for satisfactory dilution of all contaminants of concern. Table 7.2 can be used as a first estimate. If a mass balance model is used to calculate dilution air rate, imperfect mixing of dilution air must be considered.

2. Locate exhaust openings near sources of contaminants.

3. All dilution air must pass through the zone where contamination is present.

4. Replace exhaust air with make-up air of acceptable quality. The pressure drop in a ventilation system frequently requires air movers on both exhaust and make-up air for proper operation.

5. The general air movement in a space should move contaminants from point sources away from people in the room.

6. Avoid reentrance of the exhaust air by discharging exhaust above the roof line away from openings and air intakes (Wilson, 1979; Drivas et al., 1972; Reible and Shair, 1981).

In industrial applications, dilution ventilation is also frequently used for explosion control and heat control (ACGIH, 1980).

Three commonly used control strategies are source control, dilution control, and removal control. Given steady-state, completely mixed conditions with no recirculation [see Equation (6.3) with $t \to \infty$], the relationship between these control strategies can be expressed as

$$C_{i,ss} = \frac{q_0(1 - F) + q_2}{q} \; C_0 + \frac{S - R}{q} \qquad (7.1)$$

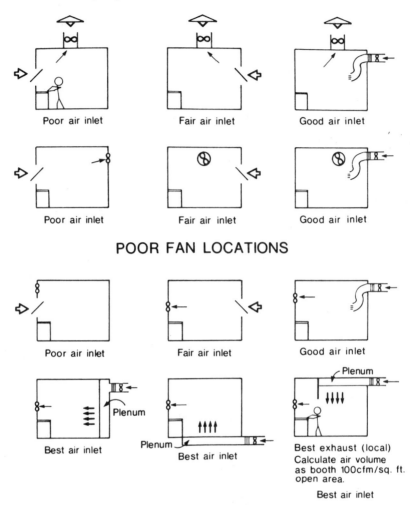

POOR FAN LOCATIONS

GOOD FAN LOCATIONS

Note: inlet air requires tempering during winter months.

Figure 7.3.   Inlet and exhaust ventilation air and fan locations [ACGIH (1980). Reprinted by permission of the Committee on Industrial Ventilation, ACGIH, Lansing, Michigan.]

where $C_{i,ss}$ = steady-state indoor concentration
C_0 = outdoor concentration
S = net generation rate
R = net removal rate
q = air flow rate (make-up air q_0 + infiltration q_2)
F = control device efficiency

Source control minimizes S by isolation of the source (e.g., no smoking) and local exhaust (e.g., a range hood). Dilution control can affect q by infiltration, natural ventilation, and forced ventilation systems. Many indoor forced-air systems recirculate 100% of the supply air. In this case, q will equal infiltration only. This can be 0.5 air changes/hr or less in a new, energy-efficient home (Woods et al., 1981). Removal control affects the term containing F. This can be particle removal by a filter or electrostatic precipitator or gas removal by a sorbent. Figure 7.4 schematically illustrates the relationships between local ventilation for source control, infiltration, forced-air ventilation, and removal control (ASHRAE, 1980). Note the variety of locations where air cleaners can be located for removal control.

Special consideration must be given when incorporating recirculated air in a dilution ventilation system. The main rationale for using recirculation is the saving of heating and cooling costs. The use of recirculated air must be based on the capability to remove hazardous contaminants from the exhaust. The general policy of the ACGIH is not to recommend recirculated air if a potential contaminant can be released in sufficient quantity to cause an adverse health impact (ACGIH, 1980). The reasons for this judgment are the following:

1. Air cleaners may not be efficient enough to remove an adequate amount of the contaminant.
2. Poor maintenance of air cleaners can result in the return of a contaminant to an occupied space.
3. Improper operation or failure of an air cleaner can result in elevated concentrations of a hazardous contaminant.

The ACGIH recommends an independent analysis of each situation for the applicability of recirculation. However, even with recirculation, the total contribution of outdoor air for the control of CO_2 must never be less than 5 cfm/person.

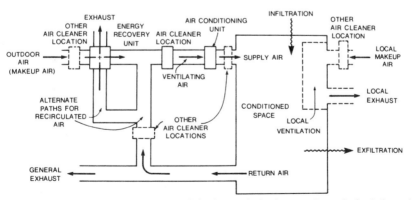

Figure 7.4. Heating, ventilation, air conditioning, and air cleaner schematic for indoor air quality control [ASHRAE (1980). Reprinted by permission of ASHRAE, New York, N.Y.]

A recent development in domestic energy conservation is the use of air-to-air heat exchangers. These energy recovery units retain indoor heat (or cold) while exchanging contaminated indoor air for outdoor air. The heat transfer surface can be metal, plastic, or asbestos paper. Inlet and exhaust streams usually have separate fans and have a cross flow orientation as is shown in Figure 7.4. Flow rates typically range between 40 and 350 m³/hr (Persily, 1981; Fisk, 1981). The effectiveness is a measure of how much of the energy from room air exiting through the exchanger can be transferred to the incoming stream with 1.0 being the maximum value. Effectiveness for domestic-sized units has been reported to vary from 0.5 to 0.85 (Persily, 1981; Fisk, 1981). Energy savings from operation of such exchangers need to be discounted by the amount of energy used by the fans.

PARTICLE REMOVAL DEVICES

Filters

Cellulose, fabric, and glass-fiber filters are the devices most often used for particle removal. The type of filter to use depends on the particular applications under consideration. Mass collection efficiency, collection efficiency for submicron particles, and pressure drop are all factors in filter evaluation. ASHRAE has recommended several standard tests for evaluating filter performance (ASHRAE, 1979). *Weight arrestance* is the mass collection efficiency of a filter for a standard synthetic dust. However, many particles which may cause soiling and/or be physiologically harmful are smaller than those in the standard synthetic dust, often in the submicron size range. The *DOP penetration test* (di-octyl phthalate) measures the percent penetration (equal to 100 minus percent efficiency) of 0.3-μm DOP particles through the filter. The *dust spot efficiency test* measures the rate of light change in glass fiber sampling filters upstream and downstream of the filter being tested. Untreated atmospheric dust is used, and the resulting filter efficiency reflects the efficacy of the filter in preventing wall discoloration. The *dust holding capacity test* measures the integrated amount of dust held on the filter up to the time the test was terminated, either at the maximum pressure drop specified by the manufacturer, or when two consecutive measures of arrestance fall below 85% of maximum mass collection efficiency. Table 7.5 indicates performance characteristics for some typical dry filter media, and Figure 7.5 indicates approximate relationships between ASHRAE test measures and various applications (ASHRAE, 1979). Other types of filters include viscous impingement filters (filter media coated with a viscous substance, such as oil, which acts as an adhesive for particles) and renewable media filters (automatically advanced, continuous rolls of dry or viscous impingement type materials).

Table 7.5 Efficiency of Particle Air Filters[a]

Filter Media Type	ASHRAE Weight Arrestance (%)	ASHRAE Atmospheric Dust Spot Efficiency (%)	MIL-STD 282 DOP Efficiency (%)	ASHRAE Dust-Holding Capacity (grams per 1700 m³/h cell)	Superficial Duct Velocity [m/sec (fpm)]	Pressure Drop mm H₂O/mm Thickness Initial	Pressure Drop mm H₂O/mm Thickness Final
Finer open cell foams and textile denier nonwovens	70–80	15–30	0	180–425			
Thin, paperlike mats of glass fibers, cellulose	80–90	20–35	0	90–180			
Mats of glass fiber multi-ply cellulose, wool felt	85–90	25–40	5–10	90–180			
Mats of 5–10 μm fibers, 6–12 mm thickness	90–95	40–60	15–25	270–540	2.54(500)	0.028[c]	0.083
Mats of 3–5 μm fibers, 6–20 mm thickness	>95	60–80	35–40	180–450			
Mats of 1–4 μm fibers, mixture of various fibers and asbestos	>95	80–90	50–55	180–360			
Mats of 0.5–2 μm fibers (usually glass fibers)	NA[b]	90–98	75–90	90–270	2.54(500)	0.054[c]	0.100
Wet laid papers of mostly glass and asbestos fibers <1 μm diameter (HEPA filters)	NA	NA	95 95–99.999 99.97	500–1000	0.64–1.3(125–250)	0.039–0.060[d]	0.174–0.255
Membrane filters (membranes of cellulose acetate, nylon, etc. having holes 1 μm diameter or less)	NA	NA	~100	NA	0.64–1.3(125–250)	0.073–0.153[d]	

[a]ASHRAE (1979).
[b]NA indicates that test method cannot be applied to this type of filter.
[c]Typical total thickness = 305 mm (12 in.).
[d]Typical total thickness = 78–300 mm (3–12 in.).

153

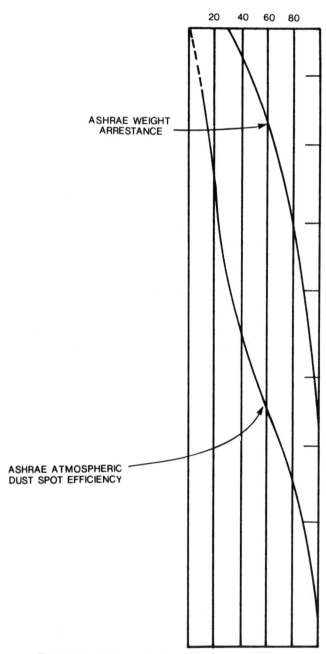

% Efficiency

20   40   60   80

ASHRAE WEIGHT
ARRESTANCE

ASHRAE ATMOSPHERIC
DUST SPOT EFFICIENCY

Typical Applications and Limitations

Window air conditioners, protection of heat exchanger from lint accumulations; relatively ineffective on smoke, settling dust, and pollen.

Window air conditioners, packaged air conditioners, domestic warm air heating; effective on lint; somewhat effective on common ragweed pollen (generally under 70%); relatively ineffective on smoke and staining particles.

Air conditioners, domestic heating, central systems; fairly effective on ragweed pollen (generally over 85%); effective as prefilter for final clean-up filters for white room, pharmaceutical, etc.; relatively ineffective on smoke and staining particles.

Same as immediately preceding but with greater degree of effectiveness; recommended minimum for make-up air for paint spray; somewhat effective in removing smoke and staining particulates.

Effective on finer airborne dust and pollen; reduce smudge and stain materially; slightly effective on fume and smoke; ineffective on tobacco smoke. Used in building recirculated and fresh air systems; some types used in domestic heating and air conditioning; used as prefilters to high-efficiency types.

Effective on all pollens, majority of particles causing smudge and stain, fume, coal and oil smoke; partially effective on tobacco smoke. Used same as above but better protection. Some types reasonably effective on bacteria.

Very effective on particles causing smudge and stain, coal and oil smoke and fume. Highly effective on bacteria. Suitable for hospital surgeries, pharmaceutical preparation areas, and controlled areas. Quite effective on tobacco smoke.

Excellent protection against bacteria, radioactive dusts, toxic dusts, all smokes and fumes. Filters in this efficiency range are generally rated by the DOP test method. Uses include hospital surgeries, intensive care wards, clean rooms, pharmaceutical packaging.

Figure 7.5. Filter applications for two ASHRAE test measures [ASHRAE (1979). Reprinted by permission of ASHRAE, New York, N.Y.]

154

In general, as arrestance or efficiency increases so will pressure drop (ΔP). Typical initial filter resistances range from 12 to 250 Pa (0.05-1.0 in. H_2O) and replacement takes place from 125 Pa (0.5 in. H_2O) for low-resistance units to 500 Pa (2 in. H_2O) for high-efficiency types such as HEPA (high-efficiency particulate air) filters. Table 7.5 lists some typical design conditions for gas velocities and initial and final ΔP's.

Electronic Air Cleaners

Electronic air cleaners use the principles of electrostatic precipitation but ordinarily operate at lower voltages than industrial installations (typically 12000 V dc potential for charging and 6000 V at the collecting plates). Electric power consumption is 12-15 W per 1700 m³/hr cleaned. Usually fitted with after and/or prefilters, the pressure drop is essentially low and constant with a range of 35-65 Pa (0.14-0.26 in. H_2O) at usual velocities of 1-2.5 m/s (ASHRAE, 1979). Figure 7.6 is a schematic of the charging and collecting electrodes. Electronic air cleaners have relatively high efficiency for submicron-sized particles such as exist in tobacco smoke. They require regular cleaning and have a higher first cost than most filter arrangements. They will also produce ozone (see Table 4.12) which may be a potential health problem if the devices are not properly installed and maintained. Other types of electronic air cleaners incorporate charged (dielectric) collection media with or without dust charging in an ionizer.

Specifications for a typical low-voltage electrostatic precipitator are given in Table 7.6. Testing of this device on welding fumes at an approach velocity of 2.79 m/s indicated that inlet concentrations in the range of 6-24 mg/m³ (much higher concentrations than usually encountered in nonindustrial settings) had

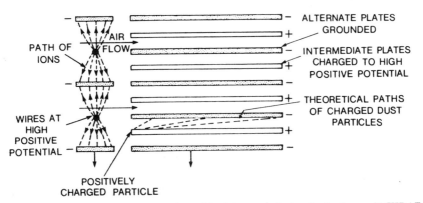

Figure 7.6. Diagrammatic cross section of ionizing-type electronic air cleaner [ASHRAE (1979). Reprinted by permission of ASHRAE, New York, N.Y.]

Table 7.6   Low-Voltage Electrostatic Precipitator Specifications[a]

Input voltage	120 V, 60 Hz, single phase
Input current	1.0 A nominal
Power consumption	100 W/0.94 m^3/s
	(100 W/2000 ACFM)
Weight	138 kg (320 lb)
Maximum recommended air capacity	0.94 m^3/s (2000 ACFM)
Ionizer section	
Voltage	8600–11,000 V dc, positive corona
Current	2.0 mA nominal
Collector section	
Plate material	Aluminum
Plate voltage	4500–5000 V dc

[a]Holcomb and Scholz (1981).

no significant effect on penetration. Two-pass operation resulted in about $\frac{1}{2}$ the penetration of one-pass operation. The average two-pass penetration for 34 runs was 8.9% for particles with a mass mean diameter of 0.6–0.7 μm (Holcomb and Scholz, 1981).

GAS FILTERS AND TRAPS

Adsorption, absorption, incineration, and catalytic conversion are all removal techniques for pollutant gases. However, adsorption is probably the most frequent recourse for control of small quantities of gaseous materials causing odor or other discomfort problems. Much of the discussion which follows was extracted from the excellent review by Amos Turk (Turk, 1977) to which the reader is referred for a more detailed evaluation.

Table 7.7 lists physical properties of a number of adsorbents. For most applications activated carbon (activated charcoal) is the adsorbent of choice as it is less selective than other materials by adsorbing organic compounds, even from humid atmospheres.

The adsorptive capacity of activated carbon can be expressed by its activity or retentivity for a reference vapor. Activity is the maximum amount of vapor which can be adsorbed when the carbon is in equilibrium with the reference vapor under specific concentration and temperature conditions. Retentivity is the maximum amount of reference material which is retained in the carbon after the vapor concentration is reduced to zero. Table 7.8 lists typical specifications for activated carbon used for air purification. The adsorption isotherms (shown in Figure 7.7) indicate maximum adsorption capacity for each substance as well as the increasing adsorptivity with molecular weight.

Table 7.7   Surface Areas and Pore Sizes of Adsorbents[a]

	Activated Carbon	Activated Alumina	Silica Gel	Molecular Sieve
Surface area (m^2/g)	1100–1600	210–360	750	–
Surface area (m^2/cm^3)	300–560	210–320	520	–
Pore volume (cm^3/g)[b]	0.80–1.20	0.29–0.37	0.40	0.27–0.38
Pore volume (cm^3/cm^3)[b]	0.40–0.42	0.29–0.33	0.28	0.22–0.30
Mean pore diameter (nm)	1.5–2.0[c]	1.8–2.0	2.2	0.3–0.9

[a]Turk (1977).
[b]Based on total of micro and macropores.
[c]Based on micropore volume (<2.5 nm diameter); macropores (>2.5 nm) not included.

It has been suggested that dividing the fractional value of activity (measured at 40° C and 32.9 mm CCl_4 in vacuo) by the density of CCl_4 (1.595 at 20° C) provides a direct estimate of micropore volume (pore diameter, $d_p \stackrel{<}{\sim} 3$ nm) (Juhola, 1975). This volume appears to be more closely related to the maximum adsorption space, W_0 (cm^3/g), than the total pore volume. Urano et al. (1982) found that W_0 was equal to 0.055 ml plus the micropore volume ($d_p < 3.2$ nm; volume determined from pore size distributions based on adsorption isotherms of nitrogen). Typical micropore volumes range from 0.3–0.5 cm^3/g.

Activated carbon can be combined with an inert carrier (e.g., paper, textiles, or extruded plastic filaments) to provide a variety of adsorption bed designs. The carrier materials contain between 50 and 80% carbon. Thin bed adsorbers

Table 7.8   Typical Specifications for Activated Carbon Used for Air Purification[a]

Property	Specification
Activity for CCl_4[b]	At least 50%
Retentivity for CCl_4[c]	At least 30%
Apparent density	At least 0.4 g/ml
Hardness (ball abrasion)[d]	At least 80%
Mesh distribution	6–14 range (Tyler sieve series)

[a]Turk (1977).
[b]Maximum saturation of carbon, at 20°C and 760 Torr in an airstream equilibrated with CCl_4 at 0°C.
[c]Maximum weight of adsorbed CCl_4 retained by carbon exposure to pure air at 20°C and 760 Torr.
[d]Percent of 6–8 mesh carbon which remains on a 14-mesh screen after shaking with 30 steel balls of 0.25–0.37 in. (0.635–0.940 cm) per 50 g carbon, for 30 min in a vibrating or tapping machine.

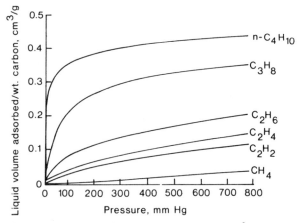

Figure 7.7. Adsorption isotherms of hydrocarbon vapors at 100°F on air purification activated carbon. Liquid volumes measured at boiling points of the hydrocarbons [Turk (1977). Reprinted by permission of A. Turk and Academic Press, New York, N.Y.]

have greater capacity for pollutant removal than do carbon impregnated designs. Flat, cylindrical, and pleated bed shapes are in use. Cylindrical canisters are designed for about 0.7 m³/min of air, pleated cells, 20–30 m³/min, and beds with aggregates of flat cells, 60 m³/min. Superficial velocities for flat bed applications are typically 0.12–0.25 m/s, and pressure drop characteristics are given in Figure 7.8. Thin bed thicknesses range from 1.3 to 1.8 cm with 6–16 mesh carbon.

Table 7.9 gives some typical, but somewhat conservative, activated carbon requirements for odor control with various occupancies (ASHRAE, 1976). However, lighter gases such as ammonia and formaldehyde will not be efficiently adsorbed with activated charcoal. A possible solution is impregnation of the adsorbent with a chemical agent that will react with a specific pollutant gas, such as sodium sulfite on activated carbon for formaldehyde (ASHRAE, 1980). Odors can be controlled by passing air through oxidizing agents such as alumina impregnated with potassium permanganate, a combination which will also remove formaldehyde (ASHRAE, 1980).

The duration of service of a particular carbon bed can be estimated by

$$t_s = \frac{2.41 \times 10^7 \, R_t W}{\eta q M C} \tag{7.2}$$

where t_s is the time of adsorbent service to saturation (hr); R_t, retentivity (fraction); W, weight of adsorbent (kg); η, adsorption efficiency (fraction); M, average molecular weight of adsorbed vapor (g/mole); q, airflow rate (m³/hr); and C, entering pollutant concentration (ppm). Table 7.10 lists retentivities and molecular weights for a variety of substances. The decision of whether to discard

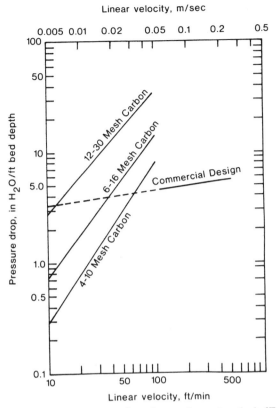

Figure 7.8. Pressure drop vs. flow rate through granular carbon beds [Turk (1977). Reprinted by permission of A. Turk and Academic Press, New York, N.Y.] – – – – – manufacturer's data, activated carbon filter, activity = 50% CCl_4 rating.

or regenerate the carbon depends on the cost and amount of the adsorbent and the length of service for a particular application.

An alternate method of estimating time to saturation for a given removal efficiency has been described by Sansone and Jonas (1981). Saturation time is characterized by a mass balance on the adsorbed vapor assuming a pseudo-first-order adsorption rate. The resulting equation, first derived by Wheeler and Robell (1969), is

$$t_s = \frac{W_e}{Cq} \left[W + \frac{\rho_B q}{k_v} \ln(1 - \eta) \right] \tag{7.3}$$

where W_e is the kinetic adsorption capacity, kg_{vapor}/kg_{carbon}, at the arbitrarily chosen η, ρ_B is the bulk density of the packed bed, kg_{carbon}/m^3, and k_v is the pseudo-first-order adsorption rate constant, hr^{-1}. The other variables are as de-

Table 7.9   Yearly Requirement in Pounds of 50-min Activated Coconut Shell Charcoal for Odor Control for Various Occupancies[a]

Residences	
Total per person	
Nonsmoking	1
Smoking	2
Individual rooms per person	
Living room	0.5
Dining room	0.25
Kitchen	0.25
Kitchen and bath	0.25
Bedroom	0.15
Laundry	0.15
Office	
Total per person	
Nonsmoking	1
Smoking	2
Hotel, per person, average occupancy	2
School, per pupil	1.5
Hospital (room or ward), per bed	3
Laboratory (average), per person	3
Bar or tavern, per occupant	5

[a]ASHRAE (1976). Add up the amount of activated charcoal needed by units as indicated in the table. For example, if there are 6 people in an office, 3 of whom are smokers and 3 nonsmokers, a total of 9 lb for one year's service under average conditions is indicated.

scribed previously and t_s will be in hours if the units of q are m^3/hr, C, kg/m^3, and W, kg of carbon. The value of η as a fraction can alternately be expressed as $(C - C_e)/C$, where C_e is the vapor concentration that penetrates the bed. The values of W_e and k_v for a particular carbon and a specific vapor can be determined experimentally from plots of t_s versus W at a selected C_e/C (Sansone and Jonas, 1981; Sansone et al., 1979; Golovoy and Braslaw, 1981b).

It has also been observed that a well-packed bed of activated carbon granules displays an adsorption capacity very close to the adsorption capacity at equilibrium (Jonas and Rehrmann, 1973). This finding provided a means for estimating W_e for a variety of compounds on a particular carbon, given kinetic capacity measurements with only one reference vapor. The Dubinin–Radushkevich equilibrium equation (Dubinin, 1975) for fine-grained carbon has then been used to extend the information:

Table 7.10   Rententivity of Vapors by Activated Carbon[a]

Substance	Formula	Molecular Weight	Normal Boiling Point (°C)	Approximate Retentivity[b] in % at 20°C 760 Torr
Acetaldehyde	C_2H_4O	44.1	21	7
Amyl acetate	$C_7H_{14}O_2$	130.2	148	34
Butyric acid	$C_4H_8O_2$	88.1	164	35
Carbon tetrachloride	CCl_4	153.8	76	45
Ethyl acetate	$C_4H_8O_2$	88.1	77	19
Ethyl mercaptan	C_2H_6S	62.1	35	23
Eucalyptole	$C_{10}H_{18}O$	154.2	176	20
Ozone	O_3	48.0	−112	Decomposes to oxygen
Phenol	C_6H_6O	94.1	182	30
Putrescine	$C_4H_{12}N_2$	88.2	158	25
Skatole	C_9H_9N	131.2	266	25
Toluene	C_7H_8	92.1	111	29
Valeric acid	$C_5H_{10}O_2$	102.1	187	35

[a]Turk (1977).
[b]Percent retained in a dry airstream at 20°C, 760 torr, based on weight of carbon.

$$\ln W_e = \ln (W_0 \rho_L) - \kappa \frac{[RT \ln (P_0/P)]^2}{\beta^2} \qquad (7.4)$$

where W_0 is the maximum adsorption space for condensed adsorbate, cm³/g; κ a structural constant of the adsorbent, (mol/cal)²; P_0 the saturated vapor pressure of the condensed adsorbate at gas temperature T, °K; P the equilibrium pressure of adsorbate vapor at T (i.e., the partial pressure of material in the gas phase entering the bed); ρ_L the liquid adsorbate density, g/cm³; R the ideal gas constant, 1.987 cal/g mol · °K; and β the dimensionless affinity coefficient. The values of κ and W_0 are parameters of a particular carbon and independent of the vapor adsorbed. The affinity coefficient can be estimated from

$$\beta = \frac{[(n_i^2 - 1)/(n_i^2 + 2)]\, M_i/\rho_{L_i}}{[(n_r^2 - 1)/(n_r^2 + 2)]\, M_r/\rho_{L_r}} \qquad (7.5)$$

where n is the refractive index determined at the sodium D wavelength; M the molecular weight; ρ_L the liquid density; and the subscripts i and r refer to a particular compound and the reference vapor, respectively (Reucroft et al., 1971).

With a known κ and experimentally determined value of W_e for the reference vapor (t_s v. W), W_0 can be determined from Eq. (7.4) for the test conditions of temperature and concentration ($\beta = 1.0$). W_e for another compound is calculated using Eqs. (7.5) and (7.4), recognizing that W_0 and κ will remain constant. k_v

for the new compound can be estimated from (Jonas et al., 1979)

$$k_{v_i} = k_{v_r} \left(\frac{M_r}{M_i}\right)^{0.5}$$ (7.6)

Equation (7.3) is then used to calculate time to saturation, or alternately, required quantity of adsorbent, for known values of ρ_B (ordinarily approximately 500 kg/m^3), C, and q, and the same value of η specified when testing the reference vapor.

Equation (7.3) is properly applicable only at high values of η. Extensive testing has been carried out at $C_e/C = 0.01$ ($\eta = 0.99$) on a Barnebey–Cheney activated carbon (6–10 mesh BC–AC, mean diameter = 0.268 cm) for carbon tetrachloride (CCl$_4$), chloroform, benzene, p-dioxane, sec-butylamine, 1,2-dichloroethane, chlorobenzene, and acrylonitrile (Sansone and Jonas, 1981; Sansone et al., 1979). Carbon tetrachloride was used as a reference vapor and the estimation procedure described above was compared to actual measurements. Calculated values agreed with experimental observations within 8% for W_e and within 15% for k_v. Based on Eqs. (7.4), and (7.6), W_e and k_v were estimated for $\eta = 0.99$ for a variety of carcinogenic materials and the specific carbon (Table 7.11) (Sansone and Jonas, 1981).

Testing of paint solvents (n-butanol, m-xylene, cellosolve acetate, and 2-heptanone) on Columbia JXC activated carbon (0.37 cm average particle size, 60% CCl$_4$ activity, minimum) gave good agreement with Eq. (7.3) at all $\eta \geqslant 0.90$ and matched reasonably well for $\eta > 0.5$ (Golovoy and Braslaw, 1981b). In addition, equilibrium and kinetic capacity measurements agreed within 14% which supported observations made by others. Estimated values of k_v using the physical and transport properties of the solvent in the gas stream were within 14% of the experimental observations. The values of k_v were 2930 min^{-1} for butanol, 3320 min^{-1} for m-xylene, 2730 min^{-1} for cellosolve acetate, and 2960 min^{-1} for heptanone. In a related study of equilibrium conditions with the same carbon, adsorption capacities were determined for n-heptane, toluene, m-xylene, isopropanol, isobutanol, n-butanol, cellosolve, ethyl acetate, cellosolve acetate, n-butyl acetate, acetone, methyl ethyl ketone, methyl isobutyl ketone, and methyl amyl ketone (Golovoy and Braslaw, 1981a). Normal heptane, n-butanol, and methyl isobutyl ketone were compared as reference materials for all the substances. Each of the reference vapors gave reasonable predictions with Eq. (7.4) and (7.5) although the least variation was found with n-heptane.

Equilibrium values of W_e for 13 vapors on 7 commercial Japanese granular activated carbons were compared with estimates from Eq. (7.4) using benzene as a reference material (Urano et al., 1981). The measured and predicted values agreed within 2% for the carbons designated B and E in Table 7.12. The com-

Table 7.11 Predicted Values of W_e and k_v for Carcinogenic Vapors on 6-10 Mesh BC–AC Activated Carbon[a]

Vapor	Liquid Density[b] ρ_L (g/cm^3)	Refractive Index[c] n	k_v (min^{-1})	W_e (kg/kg)
Acetamide	1.159	1.4274[78]	1186	0.494
Acrylonitrile	0.8060	1.3911	1251	0.357
Benzene	0.8761[23]	1.5011	1031	0.409
Carbon tetrachloride	1.5881[23]	1.4607	735	0.741
Chloroform	1.4832	1.4459	834	0.688
bis(Chloromethyl) ether	1.315	1.4346	850	0.608
Chloromethyl methyl ether	1.0605	1.3974	1016	0.480
1,2-Dibromo-3-chloropropane	2.093[11]	1.553[11]	593	0.992
1,1-Dibromoethane	2.0555	1.5128	665	0.962
1,2-Dibromoethane	2.1792	1.5383	665	1.020
1,2-Dichloroethane	1.2492[23]	1.4448	916	0.575
Diepoxy butane (meso)	1.1157	1.4330	982	0.510
1,1-Dimethyl hydrazine	0.791[22]	1.4075[22.3]	1176	0.359
1,2-Dimethyl hydrazine	0.8274	1.4204	1176	0.375
Dimethyl sulfate	1.332	1.3874	812	0.615
p-Dioxane	1.0333[23]	1.4220	971	0.475
Ethylenimine	0.8321	1.412[25]	1388	0.354
Hydrazine	1.0083	1.4698[22.3]	1610	0.380
Methyl methane sulfonate	1.2943	1.4140	869	0.595
1-Naphthylamine	1.229[25]	1.6703[51]	762	0.585
2-Naphthylamine	1.0614[98]	1.6493[98]	762	0.506
N-Nitrosodiethylamine	0.9422	1.4386	902	0.442
N-Nitrosodimethylamine	1.0059	1.4368	1059	0.458
N-Nitroso-N-methylurethane	1.133	1.4363	793	0.534
N-Nitrosopiperidine	1.0631[18.5]	1.4933[18.5]	853	0.501
N-Nitrosodipropylamine	0.9163	1.4437	799	0.434
1,3-Propane sultone	1.393[10]	1.450[10]	825	0.646
β-Propiolactone	1.1460	1.4118	1074	0.508
Propylenimine	0.802[25]	1.409[25]	1206	0.361
Safrole	1.096	1.5383	716	0.522
Urethane	0.9862[21]	1.4144[52]	966	0.456
Vinyl chloride	0.9114	1.3700	1153	0.404

[a]Barnebey–Cheney Co., lot 0993; $T = 23°C$, $P/P_0 = 0.0936$, $W_0 = 0.481$ cm^3/g, $\kappa = 1.5 \cdot 10^{-8}$ (mole/cal)^2: Table from Sansone and Jonas (1981).
[b]Liquid densities at 20°C relative to water unless otherwise stated.
[c]Refractive indices are for D line of the sodium spectrum and at 20°C unless otherwise stated.

Table 7.12   Adsorption Parameters for Various Commercial Activated Carbons

Description	Surface Area (m^2/g)	Packed Density (g/cm^3)	W_0^a (cm^3/g)	κ $(mol/cal)^2 \times 10^8$
Urano et al. (1982); 25°C				
A-Tsurumi 4GS-S, from coal and coconut shell	1170	0.43	0.485	5.1
B-Tsurumi HC-8, from coconut shell	1270	0.44	0.525	4.7
C-Takeda Sx, from coconut shell	1090	0.41	0.455	4.2
D-Hokuetsu Y-20 from coconut shell	1098	0.45	0.440	5.1
E-Fujisawa D-CG, from coconut shell	1240	0.43	0.495	5.2
F-Fujisawa A, from coal	840	0.42	0.375	4.6
G-Kureha G-BAC, from oil pitch	1000	0.51	0.430	4.6
Golovoy and Braslaw (1981a,b)				
H-Columbia JXC 4/6 pellets; 25°C, 737 mm	1194	0.461	0.404^b 0.413^c 0.409^d	1.44^b 4.7^c 1.78^d
Sansone and Jonas (1981)				
I-Barnebey–Cheney 6-10 mesh BC-AC from lot 0993; 23°C	–	–	0.481	1.5

^{a}Reference vapors: benzene for A-G; n-heptane, n-butanol and MIBK for H; CCl_4 for I.
^{b}n-heptane.
^{c}n-butanol.
^{d}MIBK.

pounds tested included toluene, o-xylene, nitrobenzene, methyl alcohol, ethyl alcohol, formic acid, ethyl acetate, acetone, methyl ethyl ketone, chloroform, carbon tetrachloride, and trichloroethylene.

In general, Eq. (7.2) will predict a greater amount of carbon to produce a given t_s than will Eq. (7.3), all other parameters being equal. For the carbon characteristics reported in Sansone and Jonas (1981), Golovoy and Braslaw (1981a), and Urano et al. (1982), W from Eq. (7.2) will be 1.5–2.5 times the W estimated from Eq. (7.3). This is illustrated in Table 7.13 for carbon tetrachloride and ethyl acetate. However, if one set of data for a reference vapor on a particular carbon is available, extrapolations using Eqs. (7.3)–(7.6) can be made with considerable confidence.

As an example of installed performance, values of the removal efficiency of activated charcoal for ozone in smoggy air [F_0 in Equations (6.3) and (6.4)] are given in Table 7.14 (Shair, 1981). In this application the adsorbent would also be expected to collect a variety of hydrocarbons present in southern California air.

Table 7.13 Comparison of Estimated Time to Saturation and Amount of Carbon[a]

Carbon from Table 7.12 Eq. (7.3)	t_s/W (hr/kg) Carbon Tetrachloride	Ethyl Acetate
A	525	575
B	573	628
C	501	552
D	476	521
E	535	584
F	410	451
G	470	517
H	460	513
I	556	627
Eq. (7.2)	339	287

[a]Calculations based on operating conditions reported in Sansone and Jonas (1981): $P/P_0 = 0.0936$; $T = 296°K$; $\eta = 0.99$; $q = 285$ cm^3/min; for CCl_4, $P_0 = 100$ mm Hg, $\rho_L = 1.588$ g/cm^3, $M = 153.8$, $n = 1.4607$, $C = 0.0123 \times 10^6$ ppm (or 7.79×10^{-8} kg/cm^3); for ethyl acetate, $P_0 = 86.6$ mm Hg, $\rho_L = 0.8974$ g/cm^3, $M = 88$, $n = 1.37$, $C = 0.0107 \times 10^6$ ppm (or 3.86×10^{-8} kg/cm^3); values of W_0 and κ taken from Table 7.12 and W_e from Table 7.11 or estimated from Eqs. (7.4) and (7.5); for carbons A-G, benzene was the reference vapor, n-heptane for carbon H, CCl_4 for carbon I; $R = 1.987$ cal/(g · mol °K); R_t from Table 7.10. For the above conditions the second term on the right-hand side of Eq. (7.3) can be neglected for W greater than about 50 g. k_v can be calculated from Eq. (7.6). For carbon I, $k_{v_r} = 735$ min^{-1} from Table 7.11; k_{v_r} was not measured for carbons A-G.

Table 7.14 Efficiency of Activated Charcoal for Ozone Collection from Smoggy Air[a]

Time Since Installation (hr)	Removal Efficiency (%) F_0	Activity
0	95 ± 5	
1200	95 ± 5	64% of initial value
2400	80	
3600	50	

[a]Shair (1981). Approximately 14,000 cfm; Pasadena, California, May–October operation. Approximately 1000 kg charcoal, fitted with 30/30 particulate prefilters.

AIR CONDITIONING

Although air conditioning serves primarily as a heat control device, it does exert some effect on indoor air quality. Pollen concentrations are significantly reduced by air conditioning (Yocum et al., 1977). Other particulate matter appears to be slightly reduced. For instance, Dockery and Spengler (1981) found that air conditioning had a weak effect on respirable particle infiltration ($p < 0.20$) and on particulate sulfate infiltration ($p < 0.05$). However, their study suggested that the effect of cigarette smoke particles is more persistent with air conditioning, due to recirculation, and was apparently more important for indoor concentrations than increased filtering of outside air. Reactive gases such as SO_2 may be reduced because of absorption by condensed moisture on heat exchanger coils, but CO tends not to be affected (Yocum et al., 1977). As has been mentioned in Chapters 2 and 4, air conditioning can also serve to concentrate infectious organisms present in outdoor air (NAS, 1981).

REFERENCES

ACGIH (1980). *Industrial Ventilation, 16th ed.* American Conference of Governmental Industrial Hygienists.

ASHRAE (1973). *ASHRAE Standard 62-73: Standards for Natural and Mechanical Ventilation.* American Society for Heating, Refrigerating and Air-Conditioning Engineers, New York.

ASHRAE (1975). *ASHRAE Standard 90-75: Energy Conservation in New Building Design.* American Society for Heating, Refrigerating and Air-Conditioning Engineers, New York.

ASHRAE (1976). *ASHRAE Handbook. 1976 Systems.* American Society of Heating, Refrigerating, and Air-Conditioning Engineers, New York.

ASHRAE (1979). *ASHRAE Handbook and Product Directory. 1979 Equipment.* American Society of Heating, Refrigerating, and Air-Conditioning Engineers, New York.

ASHRAE (1980). *Standards for Ventilation Required for Minimum Acceptable Indoor Air Quality: ASHRAE 62-73R.* American Society of Heating, Refrigerating, and Air-Conditioning Engineers, New York.

Dockery, D. W., and Spengler, J. D. (1981). Indoor–Outdoor relationships of respirable sulfates and particles. *Atmos. Environ.* 15:335–343.

Drivas, P. J., Simmonds, P. G., and Shair, F. H. (1972). Experimental characterization of ventilation systems in buildings. *Environ. Sci. Technol.* 6: 609–614.

Dubinin, M. M. (1975). Physical adsorption of gases and vapors in micropores. *Prog. Surf. Membr. Sci.* 9:1–70.

Fisk, W. J., Archer, K. M., Boonchanta, P. and Hollowell, C. D. (1981). Performance measurements for residential air-to-air heat exchangers. Lawrence Berkeley Laboratory, University of California. International Symposium on Indoor Air Pollution, Health and Energy Conservation, University of Massachusetts, Amherst, Massachusetts, October.

Golovoy, A. and Braslaw, J. (1981a). Adsorption of automotive paint solvents on activated carbon: I. Equilibrium adsorption of single vapors. *J. Air Pollut. Control Assoc.* **31**: 861–865.

Golovoy, A. and Braslaw, J. (1981a). Adsorption of automotive paint solvents on activated carbon: II. Adsorption kinetics of single vapors. National Meeting of the American Institute of Chemical Engineers, New Orleans, November.

Hellman, J. M. and Small, F. H. (1974). Characterization of the odor properties of 101 petrochemicals using sensory methods. *J. Air Pollut. Control Assoc.* **24**: 979–982.

Holcomb, M. L. and Scholz, R. C. (1981). *Evaluation of air cleaning and monitoring equipment used in recirculation systems.* NIOSH Publication No. 81–113, National Institute for Occupational Safety and Health, Cincinnati, April.

HPAC (1942). Your place in the "smart man's war." *Heating, Piping and Air Conditioning* **14**: 463 (August).

Jonas, L. A. and Rehrmann, J. A. (1973). Predictive equations in gas adsorption kinetics. *Carbon* **11**: 59–64.

Jonas, L. A., Tewari, Y. B., and Sansone, E. B. (1979). Prediction of adsorption rate constants of activated carbon for various vapors. *Carbon* **17**: 345–349.

Juhola, A. J. (1975). Thermal activation. In: *NATO Advanced Study Institute on Sorption and Filtration Methods for Gas and Water Purification*, M. Bonneive-Svendsen, Ed., Noordhoff, Leyden.

Leonardos, G., Kendall, D., and Barnard, N. (1969). Odor threshold determinations of 53 odorant chemicals. *J. Air Pollut. Control Assoc.* **19**: 91–95.

NAS (1981). *Indoor Pollutants.* National Academy of Sciences, Washington, D.C.

Olivieri, J. B. (1979). Energy conservation and comfort. *ASHRAE Trans.* **85** (Part 2): 799–808.

Persily, A. (1981). Realistic evaluation of an air-to-air exchanger. Center for Energy and Environmental Studies, Princeton University. International Symposium on Indoor Air Pollution, Health and Energy Conservation, University of Massachusetts, Amherst, Massachusetts, October.

Reible, D. D. and Shair, F. H. (1981). The reentrainment of exhausted pollutants into a building due to ventilation system imbalance. Division of Chemistry and Chemical Engineering, California Institute of Technology, Pasadena, California. International Symposium of Indoor Air Pollution, Health and Energy Conservation, University of Massachusetts, Amherst, Massachusetts, October.

Reucroft, P. J., Simpson, W. J., and Jonas, L. A. (1971). Sorption properties of activated carbon. *J. Phys. Chem.* **75**:3526–3531.

Sansone, E. B., Tewari, Y. B., and Jonas, L. A. (1979). Prediction of removal of vapors from air by adsorption on activated carbon. *Environ. Sci. Technol.* **13**:1511–1513.

Sansone, E. B. and Jonas, L. A. (1981). Prediction of activated carbon performance for carcinogenic vapors. *Am. Ind. Hyg. Assn. J.* **42**:688–691.

Shair, F. H. (1981). Relating indoor pollutant concentrations of ozone and sulfur dioxide to those outside: economic reduction of indoor ozone through selective filtration of the make-up air. *ASHRAE Trans.* **87**(Part 1):116–139.

Turk, A. (1977). Adsorption. In: *Air Pollution*, A. C. Stern, Ed., Vol. IV, 3rd ed., Academic Press, New York.

Urano, K., Omori, S., and Yamamoto, E. (1982). Prediction method for adsorption capacities of commercial activated carbons in removal of organic vapors. *Environ. Sci. Technol.* **16**:10–14.

Wheeler, A. and Robell, A. J. (1969). Performance of fixed-bed catalytic reactors with poison in the feed. *J. Catal.* **13**:299–305.

Wilson, D. J. (1979). Flow patterns over flat-roofed buildings and application to exhaust stack design. *ASHRAE Trans.* **85** (Part 2):284–285.

Woods, J. E. (1980). Environmental implications of conservation and solar space heating. Energy Research Institute, Iowa State University, Ames, Iowa, BEUL 80-3. Meeting of the New York Academy of Sciences, New York, January 16.

Woods, J. E., Maldonado, E. A. B., and Reynolds, G. L. (1981). Safe and energy efficient control strategies for indoor air quality. Energy Research Institute, Iowa State University, Ames, Iowa, BEUL 81-01. Meeting of the American Association for the Advancement of Sciences, Toronto, January 3–8.

Yocum, J. E., Coté, W. A., and Benson, F. B. (1977). Effects of indoor air quality. In: *Air Pollution*, A. C. Stern, Ed., Vol. II, 3rd ed., Academic Press, New York.

PART FOUR

APPLICATIONS

8

DESIGN METHODS

The preceding chapters summarize the major aspects of presently available information on indoor pollutant health effects, sources, and control. We recognize that such a compilation will need constant adjustment with advances in scientific knowledge, changes in lifestyles and energy costs, and alteration in society's conception of what problems are of greatest concern.

However, there is still a reasonable basis for characterizing and understanding the indoor environment. The completely mixed model discussed in Chapter 6 appears, ordinarily, to offer a reasonable estimate of indoor concentrations, given appropriate descriptions of source strengths and infiltration. Based on numerous studies and data reviews, the traditional value of 1 air change/h for living-space applications appears to be an appropriate value (see section on *Infiltration Estimation*). And the reduction of this rate by 50-70% through tight construction is borne out by observation. Source emission factors are less readily available, although reasonably good information exists for major gas- and kerosene-fueled appliances (Chapter 4). The mass discharge of formaldehyde, organic materials, and radon-daughters from structural elements, and the variation in these emission rates with time, requires further definition, particularly in actual installations. However, the values suggested in Chapter 4 are useful preliminary estimates.

The concentrations predicted with the mass balance model always need to be interpreted with respect to lifestyles and usage patterns. (A detailed example of a survey of cooking appliance utilization is given in Chapter 9.) The use of gas stoves for heating and other noncooking purposes has been reported by many investigators (e.g., Florey et al., 1979; Goldstein et al., 1979; Sterling and Sterling, 1979; Sterling and Kobayashi, 1981). The increase in carboxyhemoglobin due to methylene chloride in a paint stripping solution (Stewart and Hake, 1976), and the increase in respiratory irritation among people exposed to dried detergent residue from improperly diluted carpet shampoos containing sodium dodecyl sulfate (Kreiss et al., 1981), are situations which emphasize the need for such judgment.

In general then, a reasonable estimate of indoor levels is possible for many applications. What is less clear are the health implications of these exposures. The contribution of unvented gas stoves to childhood disease and decreases in lung function is still being defined (Table 2.5). However, some of the existing evidence suggests adverse effects, presumably due to nitrogen oxides, at concentration levels considerably lower than those encountered in occupational or environmental settings. If these observations are confirmed by further investigation the desirability of local ventilation control during gas cooking is certainly established. That such control can be effective is demonstrated by Figure 4.1.

The immediate health implications of involuntary smoking, particularly for infants, seem clearer than those due to gas stoves (Table 2.7). The long-term effects and carcinogenic potential of sidestream smoke is less understood and may well be a strong function of cultural patterns and lifestyle. Formaldehyde sensitization and ensuing health effects are even less susceptible to quantitation (see Chapter 2), although there is no question that formaldehyde exposures from particle board, laminates, and foam insulation have produced significant discomfort. The effect of various organic materials discharged from consumer products and the long-term effects of slightly elevated CO_2 concentrations (0.7–1.0%) which might be produced because of tight housing are other problems requiring further consideration.

Control procedures such as local venting and particle control are easily available. Techniques and equipment for controlling gaseous emissions in houses and apartments are less accessible. Air-to-air heat exchangers which allow replacement of polluted indoor air but retain 50–70% of its energy (e.g., Persily, 1981; Sauer and Howell, 1981; Gudac et al., 1981; Fisk et al., 1981) will be of some help in this regard. Other energy-saving procedures, such as use of a restrictor to reduce the open area of the furnace vent to the chimney, need to be evaluated for their pollution potential. The latter device increases the pressure drop through the furnace system which could conceivably lead to greater leakage of NO_x and CO combustion products into living spaces.

The lack of regulatory definition and responsibility and the absence of indoor air quality standards in the United States have been discussed in Chapter 1. This void presents the designer and evaluator with the additional problem of interpreting the health effects information to determine whether an indoor environment is satisfactory. ASHRAE 62-73R is useful in at least supplying some design guidelines (Chapter 7). However, it is well to recognize that the less stringent requirements of ASHRAE 90-75 are those more prevalent in present local and state building codes (NAS, 1981). In addition, as will become evident from the example applications of Chapters 9–11, the recommended ventilation rates of ASHRAE 63-73R need to be interpreted judiciously.

The example applications which follow illustrate how the information contained in Chapters 1–7 may be used. We believe the problems and solutions are realistic but would emphasize that any calculated estimate needs to be interpreted in the light of experience and common sense.

REFERENCES

Fisk, W. H., Archer, K. M., Boonchanta, P., and Hollowell, C. D. (1981). Performance measurements for residential air-to-air heat exchangers. Lawrence Berkeley Laboratory, University of California. International Symposium on Indoor Air Pollution, Health and Energy Conservation, University of Massachusetts, Amherst, Massachusetts, October.

Florey, C. duV., Melia, R. J. W., Chinn, S., Goldstein, B. D., Brooks, A. G. F., John, H. H., Craighead, I. G., and Webster, X. (1979). The relation between respiratory illness in primary school children and the use of gas for cooking. III—Nitrogen dioxide, respiratory illness and lung infection. *Int. J. Epid.* 8: 347–353.

Goldstein, B. D., Melia, R. J. W., Chinn, S., Florey, C. duV., Clark, D., and John, .H. H. (1979). The relation between respiratory disease in primary school children and the use of gas for cooking. II—Factors effecting nitrogen dioxide levels in the house. *Int. J. Epid.* 8:339–340.

Gudac, G. J., Mueller, M. A., Bosch, J. J., Howell, R. H., and Sauer, H. J. (1981). Effectiveness and pressure drop characteristics of various types of air-to-air energy recovery systems. *ASHRAE Trans.* 87 (Part 1), 199–210.

Kreiss, K., Gonzalez, M. G., Conright, K. L., and Scheere, A. R. (1981). Respiratory irritation due to carpet shampoo: Two outbreaks. Centers for Disease Control, Atlanta, Colorado Department of Health, Denver, and Tri-County District Health Department, Englewood, Colorado. International Symposium on Indoor Air Pollution, Health and Energy Conservation, University of Massachusetts, Amherst, Massachusetts, October.

Persily, A. (1981). Realistic evaluation of an air-to-air exchanger. Center for Energy and Environmental Studies, Princeton University. International Symposium on Indoor Air Pollution, Health and Energy Conservation, University of Massachusetts, Amherst, Massachusetts, October.

Sauer, H. J. and Howell, R. H. (1981). Promise and potential of air-to-air energy recovery systems. *ASHRAE Trans.* 87 (Part 1), 167–182.

Sterling, T. D. and Sterling, E. (1979). Carbon monoxide in kitchens and homes with gas cookers. *J. Air Pollut. Control Assoc.* 29:238–241.

Sterling, T. D. and Kobayashi, D. (1981). Use of gas ranges for cooking and heating in urban dwellings. *J. Air Pollut. Control Assoc.* 31:162–165.

Stewart, R. D. and Hake, C. L. (1976). Paint-remover hazard. *J. Am. Med. Assoc.* 235:398–401.

NAS (1981). *Indoor Pollutants.* National Academy of Sciences, Washington, D.C.

9

COMPARISON OF ASHRAE DESIGN CRITERIA FOR A RESTAURANT KITCHEN

As has been discussed in Chapters 2 and 4, gas stoves and ovens are sources of CO, NO, and NO_2. The specification of an appropriate ventilation rate for a restaurant kitchen is an application which requires consideration of the emissions from a variety of such equipment. While this situation is ordinarily more severe than for residential kitchens, the procedures are essentially the same for both applications. These include a complete inventory of combustion sources as well as a record, or estimate, of typical usage.

A number of design criteria are available for commercial kitchen design. These include the specification of 5 liters/sec · person from Table 7.2, the recommendation of 8-12 air changes/hr for food preparation areas (ASHRAE, 1979), and the general health-related guidelines of maintaining CO_2 concentrations at $\leqslant 0.5\%$ (ASHRAE, 1980a). Another limitation is the suggested air quality standards listed in Tables 1.2, 1.3, and 7.1. As will become evident, these criteria have significantly different implications for indoor air quality.

BACKGROUND DATA

Restaurant kitchen (Whitelaw, 1981)

Pollutants of interest: NO, NO_2, CO, CO_2

Location: Chicago

Outdoor levels: C_o (Table 3.1); NO_2, 63 $\mu g/m^3$; CO, 2.9 mg/m^3; CO_2 *, 0.06% (1.08 g/m^3); NO,† 78 $\mu g/m^3$.

At beginning of day (8:00 a.m.), assume $C_s = C_o$.

---

*Typical urban level.
†EPA, 1973. Ratio of NO/NO_2 on annual average basis for 1967–1971 was 1.24; determined NO by multiplying NO_2 mean value by 1.24.

Volume: Two contiguous rooms with no door between

$$22 \text{ ft} \times 15 \text{ ft} \times 10 \text{ ft} = 3300 \text{ ft}^3 \qquad 94.4 \text{ m}^3$$
$$10 \text{ ft} \times 5 \text{ ft} \times 10 \text{ ft} = \underline{500 \text{ ft}^3} \qquad \underline{14.2 \text{ m}^3}$$
$$\text{Total volume:} \qquad 3800 \text{ ft}^3 \qquad 107.6 \text{ m}^3$$

Occupancy (working hours, 8 a.m. to 11 p.m.):

Large room	2 persons	8:00 a.m.- 4:00 p.m.
	3 persons	4:00 p.m.-11:00 p.m.
Small room	1-2 persons	8:00 a.m.-11:00 p.m.

Emission sources: Based on an activity level of 2 mets/person (Table 7.3) and 0.63 ft^3 CO_2/hr · met (Table 4.16), we obtain

$$CO_2 \text{ (emitted)} = \left(\frac{2 \text{ met}}{\text{person}}\right)\left(\frac{0.63 \text{ ft}^3 \, CO_2}{\text{hr} \cdot \text{met}}\right)\left(\frac{28.3 \text{ liters}}{\text{ft}^3}\right)$$
$$\cdot \left(\frac{g \cdot \text{mole}}{24.4 \text{ liters}}\right)\left(\frac{44 \text{ g } CO_2}{g \cdot \text{mole}}\right)$$
$$= \frac{64 \text{ g}}{\text{hr} \cdot \text{person}}$$

Combustion sources and emission rates are listed in Table 9.1.

MODEL

Consider two rooms of kitchen as a single space. Ordinarily, air will not be recirculated and air from outside will not be cleaned for any of pollutants of interest. Therefore, $q_1 = 0$; $F_0 = 0$. Neglect deposition or reaction term: $R = 0$. Assume $k = 0.3$.

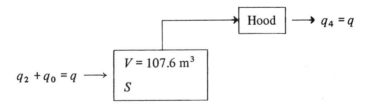

From Equation (6.3),

$$C_i = C_s e^{-(kq/V)t} + \left(\frac{S}{kq} + C_o\right)[1 - e^{-(kq/V)t}] \tag{9.1}$$

Table 9.1 Combustion Sources

Location	Usage	Heat Rate[a] (kcal/hr/unit)	Emissions (μg/kcal)[a]			
			NO	NO_2	CO	CO_2
In large room						
Gas stove						
6-(big) burners	10:30 a.m.–1:00 p.m. 5:00 p.m.–9:00 p.m. } 5–6 burners operating Rest of time—1 burner	3000	100	60	500	200,000
Burners' pilot light	8:00 a.m.–11:00 p.m.	50	40	25	200	200,000
Oven	3 times/week (for pies) between 2 and 4 p.m.	2000	60	50	500	200,000
Oven pilot light		50	2	50	1000	200,000
Broiler	4:00–5:00 p.m.	2000	60	50	500	200,000
Gas deep fryer	8:00 a.m.–11:00 p.m.	2000	60	50	500	200,000
Gas convection oven	8:00 a.m.–11:00 p.m.; 450°F	2000	60	50	500	200,000
Pilot light		50	2	50	1000	200,000
Gas convection grill	8:00 a.m.–11:00 p.m.; 2 burners	2000	60	50	500	200,000
Pilot light		50	2	50	1000	200,000
Two gas steam tables	10:30 a.m.–11:00 p.m.	50	2	50	1000	200,000
In small room						
Gas expresso machine	8:00 a.m.–11:00 p.m.	50	40	25	200	200,000

[a]From Table 4.1.

EMISSION INVENTORY

A complete emission inventory is shown in Table 9.2.

ASHRAE DESIGN CRITERIA

The following list provides the ASHRAE design criteria:

1. Recommended q for kitchen inhabitants (Table 7.2) = 5 liters/sec · person. Assume 4 persons, then

$$q = (5 \text{ liters/sec} \cdot \text{person}) (10^{-3} \text{ m}^3/\text{liter}) (3600 \text{ sec/hr}) (4 \text{ persons})$$

$$= 72 \text{ m}^3/\text{hr}$$

and the design equation [Equation (9.1)] becomes

$$C_i = C_s e^{-[((0.3)(72) t)/107.6]} + \left[\frac{S}{(0.3)(72)} + C_o \right] [1 - e^{-[((0.3)(72) t)/107.6]}]$$

(9.2)

2. Another ASHRAE design guideline for commercial kitchen ventilation is 8–12 air changes/hr (ASHRAE, 1979). For this criterion,

$$q = 12V = (12)(107.6) = 1291 \text{ m}^3/\text{hr}$$

and the design equation is

$$C_i = C_s e^{-(0.3)(12)t} + \left[\frac{S}{(0.3)(1291)} + C_o \right] [1 - e^{-(0.3)(12)t}]$$

(9.3)

3. Finally, a general health-related design guideline is that CO_2 levels remain $\leq 0.5\%$ (ASHRAE, 1980a) (see CO_2 discussion in Chapter 2). This requires that $C_i \leq 9 \text{ g/m}^3$. From the emission inventory the 5–9 p.m. period corresponds to the highest CO_2 emission rate ($S = 4930 \text{ g/hr}$ for $t = 4$ hr). Substituting in Equation (9.1), we obtain

$$C_s e^{-[(0.3)(4)/107.6] q} + \left[\frac{4930}{0.3q} + 1.08 \right] [1 - e^{-[(0.3)(4)/107.6] q}] = 9 \quad (9.4)$$

DESIGN CALCULATIONS

1. Concentrations of NO, NO_2, CO, and CO_2 are calculated for $t = 2.5$ hr (10:30 a.m.) using Equation (9.2) and $q = 72 \text{ m}^3/\text{hr}$. These predictions are

Table 9.2 Emission Inventory[a]

Emissions (g/hr)

	8:00–10:30 a.m.				10:30 a.m.–1:00 p.m.				1:00–4:00 p.m.				4:00–5:00 p.m.				5:00–9:00 p.m.				9:00–11:00 p.m.			
	NO	NO_2	CO	CO_2	NO	NO_2	CO	CO_2	NO	NO_2	CO	CO_2	NO	NO_2	CO	CO_2	NO	NO_2	CO	CO_2	NO	NO_2	CO	CO_2
Gas stove																								
Burners	0.3	0.18	1.5	600	1.5	0.9	7.5	3000	0.3	0.18	1.5	600	0.3	0.18	1.5	600	1.5	0.9	7.5	3000	0.3	0.18	1.5	600
						(5 burners)												(5 burners)						
Oven	—	—	—	—	—	—	—	—	—	—	—	—	—	—	—	—	—	—	—	—	—	—	—	—
Broiler	—	—	—	—	—	—	—	—	0.12	0.10	1.0	400	0.12	0.10	1.0	400	—	—	—	—	—	—	—	—
Pilot Lights	—	—	0.1	20	—	—	0.1	20	—	—	0.1	20	—	—	0.1	20	—	—	0.1	20	—	—	0.1	20
Deep fryer	0.12	0.10	1.0	400	0.12	0.10	1.0	400	0.12	0.10	1.0	400	0.12	0.10	1.0	400	0.12	0.10	1.0	400	0.12	0.10	1.0	400
Convection oven	0.12	0.1	1.0	400	0.12	0.1	1.	400	0.12	0.1	1.0	400	0.12	0.10	1.0	400	0.12	0.10	1.0	400	0.12	0.10	1.0	400
Pilot light	—	—	0.05	10	—	—	0.5	10	—	—	0.5	10	—	—	0.05	10	—	—	0.05	10	—	—	0.05	10
Convection grill	0.24	0.2	2.0	800	0.24	0.2	2	800	0.24	0.2	2	800	0.24	0.2	2	800	0.24	0.2	2	800	0.24	0.2	2.0	800
Pilot light	—	—	0.05	10	—	—	0.5	10	—	—	0.5	10	—	—	0.05	10	—	—	0.05	10	—	—	0.05	10
Steam tables	—	—	—	—	—	—	0.	20	—	—	0.	20	—	—	0.1	20	—	—	0.1	20	—	—	0.1	20
Expresso machine	—	—	—	10	—	—	—	10	—	—	—	10	—	—	—	10	—	—	—	10	—	—	—	10
Inhabitation	—	—	—	190	—	—	—	190	—	—	—	190	—	—	—	260	—	—	—	260	—	—	—	260

3 persons (8:00–10:30 a.m., 10:30 a.m.–1:00 p.m., 1:00–4:00 p.m.)
4 persons (4:00–5:00 p.m., 5:00–9:00 p.m., 9:00–11:00 p.m.)

	8:00–10:30 a.m.				10:30 a.m.–1:00 p.m.				1:00–4:00 p.m.				4:00–5:00 p.m.				5:00–9:00 p.m.				9:00–11:00 p.m.			
	NO	NO_2	CO	CO_2	NO	NO_2	CO	CO_2	NO	NO_2	CO	CO_2	NO	NO_2	CO	CO_2	NO	NO_2	CO	CO_2	NO	NO_2	CO	CO_2
Total Emissions, S	0.78	0.58	5.7	2440	1.98	1.30	11.80	4860	0.90	0.68	6.80	2860	0.90	0.68	6.80	2930	1.98	1.30	11.8	4930	0.78	0.58	5.8	2530

$$a_S = \frac{\text{emissions}}{\text{hour}} = \sum^{\text{No. of units}} \left(\frac{\text{heat rate}}{\text{unit} \cdot \text{hr}}\right)\left(\frac{\text{emission}}{\text{factor}}\right) + \left(\begin{array}{c}\text{Number of}\\\text{persons}\\\text{working}\end{array}\right)\left(\frac{\text{emissions } (CO_2)}{\text{person} \cdot \text{hr}}\right).$$

Only for CO_2

Table 9.3   Indoor Concentrations Using Design Equation (9.2) (Table 7.2 Criteria)

Time Period	8-10:30 a.m.	S	C_o	C_s (8 a.m.)	C_i
t	2.5 hr				
NO		0.78×10^6 $\mu g/hr$	78 $\mu g/m^3$	78 $\mu g/m^3$	$14,300$ $\mu g/m^3$
NO_2		0.58×10^6 $\mu g/hr$	63 $\mu g/m^3$	63 $\mu g/m^3$	$10,700$ $\mu g/m^3$
CO		5.7×10^3 mg/hr	2.9 mg/m^3	2.9 mg/m^3	88.8 mg/m^3
CO_2		2440 g/hr	1.08 g/m^3	1.08 g/m^3	45.6 g/m^3 (2.54%)

shown in Table 9.3. Since, for even the first time period, NO_2, CO, and CO_2 exceed acceptable health standards (see Tables 1.2, 1.3 and 7.1), $q = 72$ m³/hr is not a desirable solution.

2. Using Equation (9.3) and $q = 1291$ m³/hr, estimated values of C_i are given in Table 9.4. The results are plotted in Figure 9.1. Note that the CO_2 level of 0.5% is exceeded for 7 hr and the ambient 8-hr CO standard of 10 mg/m³ (but not the occupational standard) for 15 hr (indicated in Table 9.4 by the underline). It also is apparent that for the large value of q/V, Equation (9.3) is quite close to steady state such that

$$C_{i,ss} = \frac{S}{(0.3)(1291)} + C_o \tag{9.5}$$

3. Using the above results, it is necessary to increase q to reach a CO_2 concentration of 9 g/m³ (0.5%). Therefore Equation (9.4) simplifies to

$$\frac{4930}{0.3q} + 1.08 = 9$$

$$q = 2075 \text{ m}^3/\text{hr (20 ach)} \tag{9.6}$$

and for this criterion

$$C_{i,ss} = \frac{S}{(0.3)(2075)} + C_o \tag{9.7}$$

Using S and C_o as before, and $q = 2075$ m³/hr, results in the estimates of Table 9.5. For this case, the 8-hr environmental CO standard will still be exceeded for most of the day.

Table 9.4 Indoor Concentrations Using Design Equation (9.3) (ASHRAE Commercial Kitchen Criterion)

Time Period	8:00–10:30 a.m.	10:30 a.m.–1:00 p.m.	1:00–4:00 p.m.	4:00–5:00 p.m.	5:00–9:00 p.m.	9:00–11:00 p.m.
Time interval t (hr)	2.5	2.5	3	1	4	2
S (Table 9.2)						
NO, $\mu g/hr$	0.78×10^6	1.98×10^6	0.9×10^6	0.9×10^6	1.98×10^6	0.78×10^6
NO_2, $\mu g/hr$	0.58×10^6	1.30×10^6	0.68×10^6	0.68×10^6	1.30×10^6	0.58×10^6
CO, mg/hr	5700	11,800	6800	6800	11,800	5800
CO_2, g/hr	2440	4860	2860	2930	4930	2530
C_o						
NO, $\mu g/m^3$	78					
NO_2, $\mu g/m^3$	63					
CO, mg/m^3	2.9					
CO_2, g/m^3	1.08					
C_s						
NO, $\mu g/m^3$	78	2090	5190	2400	2400	5190
NO_2, $\mu g/m^3$	63	1560	3420	1820	1820	3420
CO, mg/m^3	2.9	17.6	33.4	20.5	20.5	33.4
CO_2, g/m^3	1.08	7.37	13.6	8.46	8.64	13.8
C_i						
NO, $\mu g/m^3$	2090	5190	2400	2400	5190	2090
NO_2, $\mu g/m^3$	1560	3420	1820	1820	3420	1560
CO, mg/m^3	17.6	33.4	20.5	20.5	33.4	17.9
CO_2, g/m^3 (%)	7.37 (0.41)	13.6 (0.75)	8.46 (0.47)	8.64 (0.48)	13.8 (0.76)	7.61 (0.42)

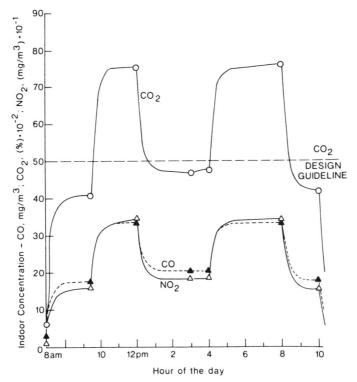

Figure 9.1. Estimated CO_2 (○), CO (▲), and NO_2 (△) concentrations in a commercial kitchen; q/V = 12 ach; V = 107.6 m^3.

LOCAL EXHAUST

As the above calculations indicate, dilution ventilation using the whole room volume sometimes is not effective if the quantity of contaminant is very great. Of course, more effective mixing with room air would lower the concentrations, but $k = 0.3$ is probably a reasonable value for this type of application. An alternate control approach is to use a local exhaust system, particularly when the emissions are concentrated in a space that is relatively small with respect to room size. Local exhaust systems ordinarily require a hood, a duct system to transport pollutants to air cleaning equipment or outside air, fan(s), exhaust stacks, and sometimes air cleaning equipment and a recirculating air system. If the hood capture velocity, the air flow rate necessary to confine pollutants to the hood, is adequate, pollutants are never allowed to mix into the room.

Although ventilating system design is somewhat beyond the scope of our discussion it is perhaps of interest to consider the implications of local exhaust

Table 9.5   Indoor Concentrations Using Equation (9.7) (CO_2 Levels $\leqslant 0.5\%$)

Time Period	8:00 a.m.–10:30 a.m.	10:30 a.m.–1:00 p.m.	1:00 p.m.–4:00 p.m.	4:00 p.m.–5:00 p.m.	5:00 p.m.–9:00 p.m.	9:00 p.m.–11:00 p.m.
NO, $\mu g/m^3$	1330	3250	1520	1320	3250	1330
NO_2, $\mu g/m^3$	990	2150	1160	1160	2150	990
CO, mg/m^3	12.1	21.9	13.8	13.8	21.9	12.2
CO_2, g/m^3(%)	5.00 (0.28)	8.89 (0.49)	5.67 (0.32)	5.79 (0.32)	9.00 (0.50)	5.14 (0.29)

for the commercial kitchen problem. The ASHRAE applications handbook (ASHRAE, 1978) indicates that 75 fpm is the lower limit for capture velocities in kitchen hoods. *Industrial Ventilation* (ACGIH, 1980) indicates guidelines for several kitchen hood designs.

A low side wall hood requires 200 cfm/lineal ft of cooking surface (ACGIH, 1980; pp. 5–109). The major combustion equipment in this kitchen can fit into an 8-ft length which results in a volumetric flow rate of 1600 cfm (45.3 m³/min). Based on the horizontal emission area being 2.5 ft $\times$ 8 ft = 20 ft² (0.57 m²), the capture velocity will be 1600/20 = 80 fpm (24.4 m/min) which is about at the lower acceptable limit. The exhaust volume corresponds to 25 ach but will essentially control pollution levels to those of ambient air. Details of hood design are given in ACGIH (1980), ASHRAE (1978), and ASHRAE (1980b).

REFERENCES

ACGIH (1980). *Industrial Ventilation*, 16th Ed. American Conference of Governmental Industrial Hygienists.

ASHRAE (1978). *ASHRAE Handbook, 1978 Applications.* American Society of Heating, Refrigerating, and Air Conditioning Engineers, New York.

ASHRAE (1979). *ASHRAE Handbook and Product Directory, 1979 Equipment.* American Society of Heating, Refrigerating and Air Conditioning Engineers, New York.

ASHRAE (1980a). *Standards for Ventilation Required for Minimum Acceptable Indoor Air Quality.* ASHRAE 62-73R, American Society of Heating, Refrigerating and Air Conditioning Engineers, New York.

ASHRAE (1980b). *ASHRAE Handbook, 1980 Systems.* American Society of Heating, Refrigerating, and Air Conditioning Engineers, New York.

EPA (1973). *The national air monitoring program: air quality and emissions trends-annual report.* Vol. 1, Chap. 4. U.S. Environmental Protection Agency Report No. EPA-450/1-73-001, Washington, D.C.

Whitelaw, K. L. (1981). (Formerly manager, Mallory's, Chicago, Ill.), personal communication.

10

AIR QUALITY IN A
CONFERENCE ROOM WITH
SMOKING: EFFECT OF
CHANGING PARAMETERS
IN THE ONE- AND
TWO-COMPARTMENT MODELS

This problem examines the predictions generated by the one- and two-compart-ment forms of the mass balance model. The one-compartment example shows the effect of varying three parameters on model predictions; deposition or re-moval; air cleaner removal efficiency, and mixing factor. The two-compartment application explores the effect of different emission patterns in the two spaces on predicted indoor air quality.

The main sources of pollution are cigarettes, outdoor air, and people. Such sources can contribute significant quantities of respirable particles, carbon monoxide, and carbon dioxide to indoor air. These three pollutants are chosen for a variety of reasons. Respirable particles (RP) are a potential health hazard, have a measured deposition velocity, and can be removed by high-efficiency filters. Carbon monoxide is also a potential health hazard and is usually studied as an index of the gas-phase components from cigarette smoke. Carbon dioxide is considered because of its importance in the ASHRAE indoor air quality guidelines.

The one-compartment mass balance model is shown in Figure 6.1. Table 10.1 gives information about the conference room size, ventilation, and use pattern.

The first step in solving a mass balance model problem is to develop an

Table 10.1   Conference Room Ventilation and Source Data

Room size	250 m^3
Ventilation rate make-up air (q_0) (0.5 ach)	75 cfm (125 m^3/hr)
recirculated air (q_1)	310 cfm (520 m^3/hr)
Filter efficiency for recirculated air (F_1)	10% for RP
	85% for dust
Average number of people in room	10
Average number of smokers in room	4
Average smoking rate per person	4 cigarettes/hr
Periods when room is used	9:00 a.m.–noon
	2:00–4:00 p.m.
Location	Chicago (downtown)

emission inventory. Table 10.2 lists characteristics of the three sources being studied. Cigarettes and respiration add pollutants to the conference room during the two periods when it is in use; 9:00 a.m.–12:00 noon and 2:00–4:00 p.m. Outdoor air continuously contributes small constant amounts of the three materials.

Equation (6.3) is used to predict indoor air quality. The following values are the initial conditions for solving this equation for RP. Note that F_1 is set equal to 0.1. Although the recirculation air filter is 85% efficient for dust, the filter has a low efficiency for the small particles in cigarette smoke. For this problem, t is in hours, and all flow rates must be on the same time basis.

Table 10.2   Source Strength Data

Source Type	Respirable Particles	Carbon Monoxide	Carbon Dioxide
Cigarettes	25.8 mg/cig[a]	37.5 mg/cig[a]	320 mg/cig[a]
	(410 mg/hr)	(0.6 g/hr)	(5.1 g/hr)
Outdoor air	0.05 mg/m^3 [b,c]	2.9 mg/m^3 [b]	0.79 g/m^3 [d]
Respiration	—	—	35 g/hr · person[e]
			(350 g/hr)

[a]Average value from Table 4.5.
[b]Table 3.1.
[c]Wadden et al., 1980.
[d]Chapter 3.
[e]ASHRAE, 1980.

BASIS FOR SOLVING THE MASS-BALANCE MODEL FOR RESPIRABLE PARTICLES IN A CONFERENCE ROOM

t in hours

q_2 (infiltration) = 0

q_3 (exfiltration) = 0

at $t = 0$ (9:00 a.m.), $C_i = C_s = C_o$

C_o = outdoor concentration

F_0 (make-up air filter) = 0

F_1 (recirculation air filter) = 0.1

R (indoor sink) = 0

S (source) = 410 mg/hr for 9:00 a.m.–12:00 noon and 2:00–4:00 p.m.;
   0 for other times

k (factor for inefficient mixing) = 0.3 (Ishizu, 1980)

Solving Equation (6.3) for RP concentration, we obtain

$$C_i = \frac{k(q_0) C_o + S}{k(q_0 + q_1 F_1)} \left[1 - e^{-(k/V)(q_0 + q_1 F_1) t} \right] + C_s e^{-(k/V)(q_0 + q_1 F_1) t}$$

$$= \frac{37.5 C_o + S}{53.10} \left[1 - e^{-0.21t} \right] + C_s e^{-0.21t} \tag{10.1}$$

The approach used throughout this problem for computing indoor concentration for a 24-hr period is as follows. Note that the equation has to be reset to $t = 0$ whenever there is a change in emission rate.

Calculational Approach

1. Solve Equation (10.1) for concentrations at 10:00, 11:00 a.m., and noon ($t = 1, 2$, and 3 hr), $C_s = C_o$.
2. From noon to 2:00 p.m., set $C_s = C_i$ at noon and assume $S = 0$ ($t = 1$ and 2).
3. From 2:00 to 4:00 p.m., set C_s to C_i at 2:00 p.m. and solve for concentration at 3:00 and 4:00 p.m. ($t = 1$ and 2).
4. From 4:00 p.m. on, set C_s to C_i at 4:00 p.m. and assume $S = 0$ ($t = 2, 6, 12$, and 16).

The solution of Equation (10.1) for RP is the following:

Time	t (hr)	C_s (mg/m^3)	S (mg/hr)	C_i (mg/m^3)
9:00 a.m.	0	0.05	–	0.05
10:00	1	0.05	410	1.51
11:00	2	0.05	410	2.69
12:00 p.m.	3	0.05	410	3.65
1:00	1	3.65	0	2.97
2:00	2	3.65	0	2.41
3:00	1	2.41	410	3.42
4:00	2	2.41	410	4.24
6:00	2	4.24	0	2.80
10:00	6	4.24	0	1.23
4:00 a.m.	12	4.24	0	0.37
8:00	16	4.24	0	0.18

CO and CO$_2$ Concentrations in the Conference Room

The procedure for solving Equation (6.3) for CO and CO$_2$ is the same as for RP except for the following changes:

$F_1 = 0$ (removal of CO and CO$_2$ at ambient levels is not practical)
$C_o(CO) = 2.9$ mg/m^3
$C_o(CO_2) = 0.79$ g/m^3
$S(CO) = 0.6$ g/hr $\Big\}$ for 9:00 a.m.–12:00 noon and 2:00–4:00 p.m.;
$S(CO_2) = 355$ g/hr $\Big\}$ 0 at other times

Solving Equation (6.3) for CO or CO$_2$ concentration, we obtain

$$C_i = \left(\frac{37.5 C_o + S}{37.5}\right)(1 - e^{-0.15t}) + C_s e^{-0.15t} \qquad (10.2)$$

For CO, we obtain

Time	t (hr)	C_s (mg/m^3)	S (mg/hr)	C_i (mg/m^3)	C_i (ppm)
9:00 a.m.	0	2.9	–	2.90	2.53
10:00	1	2.9	600	5.13	4.48
11:00	2	2.9	600	7.05	6.16

Time	t (hr)	C_s (mg/m^3)	S (mg/hr)	C_i (mg/m^3)	C_i (ppm)
12:00 p.m.	3	2.9	600	8.70	7.60
1:00	1	8.70	0	7.89	6.89
2:00	2	8.70	0	7.20	6.29
3:00	1	7.20	600	8.83	7.71
4:00	2	7.20	600	10.23	8.93
6:00	2	10.23	0	8.33	7.27
10:00	6	10.23	0	5.88	5.13
4:00 a.m.	12	10.23	0	4.11	3.59
8:00	16	10.23	0	3.56	3.11

For CO_2, we obtain

Time	t (hr)	C_s (g/m^3)	S (g/hr)	C_i (g/m^3)	C_i (%)
9:00 a.m.	0	0.79	—	0.79	0.04
10:00	1	0.79	355.1	2.11	0.12
11:00	2	0.79	355.1	3.24	0.18
12:00 p.m.	3	0.79	355.1	4.22	0.23
1:00	1	4.22	0	3.74	0.21
2:00	2	4.22	0	3.33	0.19
3:00	1	3.33	355.1	4.30	0.24
4:00	2	3.33	355.1	5.13	0.29
6:00	2	5.13	0	4.00	0.22
10:00	6	5.13	0	2.55	0.14
4:00 a.m.	12	5.13	0	1.51	0.08
8:00	16	5.13	0	1.18	0.07

Table 10.3 is a summary of peak and average concentrations of CO, CO_2, and RP. Figure 10.1 shows the predicted concentrations for RP, CO, and CO_2 as a function of time.

COMPARISON WITH ASHRAE STANDARD

Based on the pollutant mass balance model [Equation (6.2)], indoor concentrations of RP exceed the ASHRAE recommend indoor value (260 $\mu g/m^3$ for

Table 10.3   Summary of Predicted Concentrations

Pollutant	Peak Concentration	9 a.m.–4 p.m. Average Concentration	24-hr Average Concentration
RP	4.24 mg/m^3	2.62 mg/m^3	1.43 mg/m^3
CO	8.93 ppm	6.32 ppm	
CO_2	0.29%	0.19%	

24 hr). All calculations to this point assume the sink value, R, is zero. This may not be a good assumption for all pollutants. Deposition of particles less than 1-μm diameter, the size range of tobacco smoke particles, has been estimated at 0.05 hr^{-1} (Dockery and Spengler, 1981). Given a deposition rate, K, the sink term can be expressed as

$$R = KM = \text{(deposition rate) (mass)} \tag{10.3}$$

and

$$R = KVC_i = \text{(deposition rate) (room volume) (concentration)} \tag{10.4}$$

Substituting $E = KV$ and Equation (10.4) into Equation (6.2) leads to

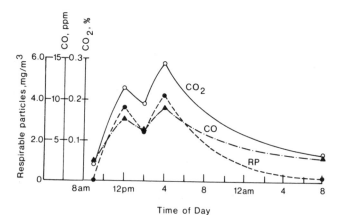

Figure 10.1. Respirable particle, CO, and CO_2 concentrations in a conference room with smoking:

	$K(\text{hr}^{-1})$	k	F_1
● RP	0	0.3	0.1
▲ CO	0	0.3	0
○ CO_2	0	0.2	0

$$C_i = C_s e^{\alpha t/V} + \frac{\beta}{\alpha} (e^{(\alpha t/V)} - 1) \qquad (6.4)$$

where

$$\alpha = -kq_1 F_1 - kq_0 - kq_2 - KV$$
$$\beta = kq_2 C_o + kq_0 (1 - F_0) C_o + S$$

Solving Equation (6.4) for RP concentrations with $K = 0.05$ hr^{-1} and all other assumptions and conditions as before, $\alpha = -65.60$ and $\beta = 1.88 + S$:

Time	t (hr)	C_S (mg/m^3)	S (mg/hr)	C_i (mg/m^3)
9:00 a.m.	0	0.05	—	0.05
10:00	1	0.05	410	1.49
11:00	2	0.05	410	2.59
12:00 p.m.	3	0.05	410	3.44
1:00	1	3.44	0	2.65
2:00	2	3.44	0	2.05
3:00	1	2.05	410	3.03
4:00	2	2.05	410	3.78
6:00	2	3.78	0	2.25
10:00	6	3.78	0	0.81
4:00 a.m.	12	3.78	0	0.19
8:00	16	3.78	0	0.09

The nonzero value of R results in lower predicted concentrations. The peak concentration at 4:00 p.m. is 0.5 mg/m^3 lower; and by the next morning the concentration has dropped by 50%. The 9 a.m.–4 p.m. average concentration is still 2.39 mg/m^3, and the 24-hr average is 1.17 mg/m^3—almost 5 times the recommended standard.

ASHRAE (1980) recommends a minimum of 5 cfm/person make-up air (q_0) for each person in a room for CO_2 control. q_0 for the conference room is 75 cfm, 25 cfm above this recommendation. Predicted CO_2 concentrations are below the ASHRAE recommended value and consequently the ASHRAE recommended ventilation rates are appropriate. Table 7.2 lists 35 cfm/person for office meeting spaces where smoking is permitted. For ten persons in the conference room, this corresponds to 350 cfm of outdoor or equivalent air. The outdoor air supply is only 75 cfm. Because of the low-efficiency filter in the recirculated air stream ($F_1 = 0.1$ for RP), the recirculated air does not meet the criteria for ventilation air. ASHRAE recommends that Equation (10.5) be used to relate F_1, q_0, and q_1:

$$q_1 = \frac{q_R - q_0}{F_1} \qquad (10.5)$$

where q_R is the recommended ventilation rate from Table 7.2 and q_0 is the actual ventilation rate of outdoor air. Given a recommended q_R of 350 cfm (600 m³/hr), q_0 of 75 cfm (125 m³/hr) and q_1 of 310 cfm (520 m³/hr), a recirculation air filter efficiency (F_1) of 0.9 will meet ASHRAE ventilation standards.

Substituting these values into Equation (6.4) and solving for the RP concentration, where $\alpha = -190.40$ and $\beta = 1.88 + S$, we obtain

Time	t (hr)	C_S (mg/m³)	S (mg/hr)	C_i (mg/m³)
9:00 a.m.	0	0.05	—	0.05
10:00	1	0.05	410	1.18
11:00	2	0.05	410	1.70
12:00 p.m.	3	0.05	410	1.95
1:00	1	1.95	0	0.91
2:00	2	1.95	0	0.43
3:00	1	0.43	410	1.36
4:00	2	0.43	410	1.79
6:00	2	1.79	0	0.40
10:00	6	1.79	0	0.03
4:00 a.m.	12	1.79	0	0.01
8:00	16	1.79	0	0.01

Using the ASHRAE recommended value for F_1, the predicted 9 a.m.–4 p.m. average RP concentration is reduced to 1.17 mg/m³, more than 50% below the previous value. However, the predicted 24-hr average concentration is 0.43 mg/m³, which is still 67% above the ASHRAE indoor recommendation.

The predicted CO concentrations (8.93 ppm peak and 6.32 ppm average) are for conditions that do not meet ASHRAE ventilation standards where smoking is allowed. Unlike particles, control devices for CO or CO_2 are less practical. The best way to meet the ASHRAE ventilation standard for CO or CO_2 is to use outside air only [$q_0 = 600$ m³/hr (350 cfm) and $q_1 = 0$]. Resolving Equation (6.4) for RP, CO, and CO_2 with $q_0 = 600$ m³/hr, $q_1 = 0$, $F_1 = 0$, and $K_{CO} = 0$, we obtain

Time	t (hr)	RP (mg/m^3)	CO (ppm)	CO_2 (%)
9:00 a.m.	0	0.05	2.53	0.04
10:00	1	1.19	4.03	0.10
11:00	2	1.72	4.75	0.13
12:00 p.m.	3	1.97	5.11	0.14
1:00	1	0.94	3.79	0.09
2:00	2	0.46	3.14	0.07
3:00	1	1.38	4.32	0.11
4:00	2	1.81	4.90	0.13
6:00	2	0.42	3.09	0.07
10:00	6	0.06	2.57	0.05
4:00 a.m.	12	0.05	2.53	0.04
8:00	16	0.05	2.53	0.04

Ventilation that meets the ASHRAE standards results in a predicted 9 a.m.–4 p.m. average CO concentration of 4.07 ppm, a reduction of 2.25 ppm from the previous conditions. This concentration is below the outdoor air standard (Table 1.2).

VALIDATION OF MODEL

RP

Table 2.6 lists concentrations of 1.1–3.0 mg/m^3 particulate matter in homes as large as 425 m^3 when up to 35 cigarettes were smoked. This is very consistent with the averages for RP calculated from Equation (6.4).

Dockery and Spengler (1981) empirically fit a model for predicting indoor respirable particulate concentration [Equations (6.20), (6.21), (6.22), and Table 6.10]. Solving this model for the conference room, we obtain

$$P\bar{C}_o = C_o\beta_1 + AC_o\beta_2$$
$$= (50.0)(0.7) - (1)(50.0)(0.39)$$
$$= 15.5 \ \mu g/m^3 \tag{6.21}$$

$$\frac{\bar{S}}{q} = \beta_3 N + \beta_4 AN + \beta_5 A + \beta_6$$

$$= (0.88)(80) + (1.23)(1)(80) - (2.39)(1) + 15.02$$

$$= 181.4 \ \mu g/m^3 \qquad\qquad (6.22)$$

$$\bar{C}_i = P\bar{C}_o + \frac{\bar{S}}{q}$$

$$= 15.5 + 181.4$$

$$= 196.9 \ \mu g/m^3 \quad \text{or} \quad 0.20 \ mg/m^3 \qquad\qquad (6.20)$$

This value assumes air conditioning and is substantially lower than the 0.43 mg/m^3 calculated with $F_1 = 0.9$ and the 1.17 mg/m^3 calculated with $F_1 = 0.1$. The empirical model, however, was developed to predict annual average concentration for homes and would be expected to predict lower concentrations than those implied by the conditions described in this problem. Consequently, the differences are not unreasonable.

CO

Table 2.6 also shows that CO concentrations ranged from 2.5 to 28 ppm in indoor environments with tobacco smoke pollution. The predicted values of 6.32 ppm (not meeting ASHRAE recommended ventilation) and 4.07 ppm (at the ASHRAE ventilation standard) are both in this range.

Effect of k on the Model

k, a factor which accounts for inefficient mixing, was estimated at 0.3. Reevaluating the model for RP with $k = 0.6$ demonstrates the effect of a more completely mixed system. For this calculation, assume make-up air rate, $q_0 = 125 \ m^3/hr$; recirculated air rate, $q_1 = 520 \ m^3/hr$; recirculated air filter efficiency, $F_1 = 0.9$ for RP and 0 for CO and CO_2; and particle deposition, $K = 0.05 \ hr^{-1}$. Substituting into Equation (6.4) and solving for RP, CO, and CO_2 we obtain

Time	t (hr)	RP (mg/m^3)	CO (ppm)	CO_2 (%)
9:00 a.m.	0	0.05	2.53	0.04
10:00	1	0.88	4.34	0.11
11:00	2	1.07	5.68	0.16

Time	t (hr)	RP (mg/m^3)	CO (ppm)	CO_2 (%)
12:00 p.m.	3	1.11	6.68	0.20
1:00	1	0.26	5.61	0.16
2:00	2	0.07	4.81	0.13
3:00	1	0.95	6.03	0.18
4:00	2	1.07	6.93	0.21
6:00	2	0.07	4.95	0.14
10:00	6	0.01	3.26	0.07
4:00 a.m.	12	0.01	2.65	0.05
8:00	16	0.01	2.57	0.04

Changing the mixing factor to 0.6 decreases the predicted 9 a.m.–4 p.m. average RP concentration to 0.68 mg/m³ and the 24-hr average concentration to 0.24 mg/m³, just below the ASHRAE recommended standard.

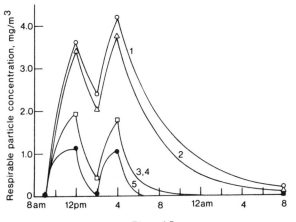

Figure 10.2. Effect of parameter variation on respirable particle concentration for a one-compartment model:

	$K(hr^{-1})$	k	F_1
○ case 1	0	0.3	0.1
△ case 2	0.05	0.3	0.1
□ cases 3 and 4	0.05	0.3	0.9
● case 5	0.05	0.6	0.9

Table 10.4 Predicted Indoor Air Quality for a Conference Room with Smoking

Pollution	q_0 (m³/hr)	q_1 (m³/hr)	k	F_1	K (hr⁻¹)	Peak Concentration	9 a.m.–4 p.m. Average Concentration	24-hr Average Concentration
RP[a]	125	520	0.3	0.1	0	4.24 mg/m³	2.62 mg/m³	1.43 mg/m³
CO	125	520	0.3	0	0	8.93 ppm	6.32 ppm	
CO₂	125	520	0.3	0	0	0.29%	0.19%	
RP[b]	125	520	0.3	0.1	0.05	3.78 mg/m³	2.39 mg/m³	1.17 mg/m³
RP[c]	125	520	0.3	0.9	0.05	1.95 mg/m³	1.17 mg/m³	0.43 mg/m³
RP[d]	600	0	0.3	0	0.05	1.97 mg/m³	1.19 mg/m³	0.46 mg/m³
CO	600	0	0.3	0	0	5.11 ppm	4.07 ppm	
CO₂	600	0	0.3	0	0	0.14%	0.10%	
RP[e]	125	520	0.6	0.9	0.05	1.11 mg/m³	0.68 mg/m³	0.24 mg/m³
CO	125	520	0.6	0	0	6.93 ppm	5.22 ppm	
CO₂	125	520	0.6	0	0	0.21%	0.15%	

[a]Case 1: Mass balance model [Equation (6.3)].

[b]Case 2: Mass balance model with deposition [Equation (6.4)]. (Note that $K_{CO} = 0$, $K_{CO_2} = 0$, and predicted concentrations of CO and CO₂ equal Case 1 values.)

[c]Case 3: q_0, q_1, and F_1 meet the ASHRAE ventilation standard for RP. The change in q_1 and F_1 (RP) will not affect predicted concentrations of CO and CO₂ because $F_{1,CO} = F_{1,CO_2} = 0$.

[d]Case 4: q_0 meets ASHRAE ventilation standards for RP, CO, and CO₂.

[e]Case 5: Case 3 conditions with $k = 0.6$.

SUMMARY OF ONE-COMPARTMENT PREDICTIONS

Figure 10.2 shows the modeled RP concentration-time relationships for the five conditions explored. Table 10.4 is a summary of predicted indoor air quality for all conditions described in this problem.

A comparison of case 1 and 2 for RP demonstrates the effect of including particle decay (through deposition) on predicted concentration. A comparison of case 2 and 3 for RP shows the effect of switching to a high-efficiency filter in the recirculated airstream. Note that case 3 meets ASHRAE recommended ventilation standards only for RP. Case 4 meets ASHRAE standards for all three pollutants through increased make-up air flow. A comparison of cases 1 and 4 for CO and CO_2 shows the result of meeting the recommended standard for a conference room with smoking. For RP, cases 1 and 2 do not meet the ASHRAE ventilation requirements and cases 3 and 4 do. Finally, comparing cases 3 and 5 demonstrates the effect of increasing the mixing factor on predicted concentration.

Particle deposition and the recirculation air particle filter efficiency have no effect on CO or CO_2 concentration. Because of this, CO and CO_2 predictions for cases 2 and 3 conditions are the same as for case 1 conditions.

ACTIVITY PATTERNS

The activity patterns of people can have a significant effect on indoor air quality. This problem has been based on occupation of a conference room for 5 hr with a 2-hr midday break. Table 10.5 compares peak concentrations at 4:00 p.m. resulting from the 5-hr schedule, and peak concentrations at 4:00 p.m. assuming occupancy for 7 straight hours. The 2-hr break in smoking and respiration has a significant effect on the 4:00 p.m. concentration for all but the well mixed RP case ($k = 0.6$).

APPLICATION OF TWO-COMPARTMENT MODEL

The two-compartment model is shown in Figure 6.11. Table 10.6 gives information about the conference room and building size, ventilation and use pattern. For this example, the two-compartment model is solved for RP concentration. Note that the conference room size, occupancy, smoking rate, use, and location are the same as for the one-compartment application. Also q_5 (conference room air) is the same flow rate as q_0 from case 4 of the one-compartment model.

Table 10.5 Comparison of 4:00 p.m. Concentrations After 5-hr and 7-hr Occupancy

Case	Pollutant	q_0 (m³/hr)	q_1 (m³/hr)	k	F_1	K (hr⁻¹)	4:00 p.m. Concentration (5-hr Occupancy)	4:00 p.m. Concentration (7-hr Occupancy)
1	RP	125	520	0.3	0.1	0	4.24 mg/m³	5.98 mg/m³
	CO	125	520	0.3	0	0	8.93 ppm	11.62 ppm
	CO_2	125	520	0.3	0	0	0.29%	0.39%
2	RP	125	520	0.3	0.1	0.05	3.78 mg/m³	5.29 mg/m³
3	RP	125	520	0.3	0.9	0.05	1.79 mg/m³	2.15 mg/m³
4	RP	600	0	0.3	0	0.05	1.81 mg/m³	2.17 mg/m³
	CO	600	0	0.3	0	0	4.90 ppm	5.42 ppm
	CO_2	600	0	0.3	0	0	0.13%	0.15%
5	RP	125	520	0.6	0.9	0.05	1.07 mg/m³	1.12 mg/m³
	CO	125	520	0.6	0	0	6.93 ppm	8.66 ppm
	CO_2	125	520	0.6	0	0	0.21%	0.27%

The sources of RP are cigarettes and outdoor air. Equations (6.17) and (6.18) are used to predict indoor concentration in the conference room and building, respectively. The constants in these equations are defined in Table 6.9. The following are the initial conditions for RP.

Table 10.6 Two-Compartment Model: Ventilation and Source Data

Room size	Conference room	250 m³
	Building	5000 m³
Ventilation rates	q_0 (building make-up air)	1000 cfm (1700 m³/hr)
	q_1 (building recirculated air)	2250 cfm (3800 m³/hr)
	q_5 (conference room air)	350 cfm (600 m³/hr)
Average number of people	Conference room	10
	Building	40
Average number of smokers	Conference room	4
	Building	16
Average smoking rate per person		4 cigarettes/hr
Periods when spaces are occupied	Conference room	9:00 a.m.–12 noon 2:00–4:00 p.m.
	Building	9:00 a.m.–4:00 p.m.
Location		Chicago (downtown)

BASIS FOR SOLVING TWO-COMPARTMENT MODEL FOR RESPIRABLE PARTICLES

t in hours

q_2 (infiltration) = 0

q_3 (exfiltration) = 0

At $t = 0$ (9:00 a.m.): $C_{1,o} = C_{2,o} = C_o$

C_o (outdoor concentration) = 0.05 mg/m^3

F_0 (make-up air filter) = 0.9

F_1 (recirculation air filter) = 0.9

S (source):

    S_1 (conference room) = 410 mg/hr for 9:00 a.m.–12 noon and 2:00–4:00 p.m.; 0 other times

    S_2 (building) = 1640 mg/hr for 9:00 a.m.–4:00 p.m.; 0 other times

k (mixing factor):

    conference room = 0.3

    building = 0.2

K (particle deposition) = 0.05 hr^{-1}

The approach used for this calculation is the same as that used for the one-compartment problem. Equations (6.17) and (6.18) along with a number of their constants, must be reset to $t = 0$ whenever there is a change in emission rates, S_1 or S_2.

The solution for RP in the conference room and RP in the building is as follows:

Time	t (hr)	S_1 (mg/hr)	S_2 (mg/hr)	RP Conference room (mg/m^3)	RP Building (mg/m^3)
9:00	0	—	—	0.05	0.05
10:00	1	410	1640	1.27	0.34
11:00	2	410	1640	1.97	0.57
12:00	3	410	1640	2.40	0.76
1:00	1	0	1640	1.26	0.70
2:00	2	0	1640	0.86	0.76
3:00	1	410	1640	1.97	0.92
4:00	2	410	1640	2.55	1.04
6:00	2	0	0	1.14	0.68
10:00	6	0	0	0.39	0.29
4:00	12	0	0	0.11	0.08
8:00	16	0	0	0.05	0.04

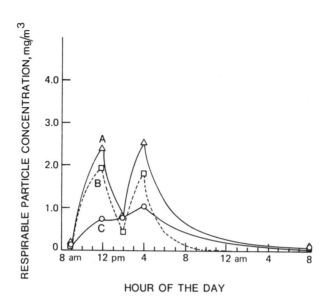

Figure 10.3. Comparison of respirable particle concentrations from one- and two-compartment models:

		$K(\text{hr}^{-1})$	k	F_1
△ A	two-compartment, internal room, $V_1 = 250\ \text{m}^3$	0.05	0.3	0.9
▢ B	one-compartment, $V = 250\ \text{m}^3$	0.05	0.3	0.9
○ C	two-compartment, external space, $V_2 = 5000\ \text{m}^3$	0.05	0.2	0.9

Table 10.7 Predicted Indoor RP Concentrations

Case	Noon (mg/m³)	4:00 p.m. (mg/m³)	9 a.m.–4 p.m. Average (mg/m³)	24-hr Average (mg/m³)
Conference room				
Two-compartment model	2.40	2.55	1.54	0.71
Building				
Two-compartment model	0.76	1.04	0.64	0.35
Conference room				
Case 4,				
One-compartment model	1.97	1.81	1.19	0.46

Table 10.7 is a summary of peak and average RP concentrations. Included in this table is case 4 from the one-compartment example. The ventilation rate ($q_0 = 600$ m^3/hr) from case 4 is equal to q_5 for the two-compartment example. Figure 10.3 shows the predicted concentration of RP in the conference room and building. Case 4 of the one-compartment model is also included in this figure. Figure 10.3 and Table 10.7 show that mixing conference room air with building air results in a higher predicted RP concentration than mixing the same volume of room air with outside air.

REFERENCES

ASHRAE (1980). *ASHRAE 62-73R: Standards for ventilation required for minimum acceptable indoor air quality*. American Society of Heating, Refrigerating, and Air Conditioning Engineers, New York.

Dockery, D. W. and Spengler, J. D. (1981). Indoor–outdoor relationships of respirable sulfates and particles. *Atmos. Environ.* **15**:335–343.

Ishizu, Y. (1980). General equation for the estimation of indoor pollution. *Environ. Sci. Technol.* **14**:1254–1257.

Wadden, R. A., Allen, R. J., Scheff, P. A., and Hogan, T. J. (1980). Aerosol size characteristics in Chicago air, in press. *Proc. 5th International Clean Air Conference.* Buenos Aires, Argentina.

11

EVALUATION
OF AIR QUALITY
INSIDE A HOME

This application is a study of air quality inside a home. In this problem, outdoor air, a gas stove, smoking, an electrostatic air cleaner, respiration, and particle board (made with urea-formaldehyde glue) contribute to, or modify, concentrations of respirable particles, CO, CO_2, NO, NO_2, HCHO, and O_3. This problem uses the pollutant mass balance model [Equation (6.2)] and the statistical models of Dockery and Spengler (1981) [Equation (6.20)] and Anderson et al. (1975) [Equation (4.4)].

PROBLEM BASIS

7-room house located in Chicago

2000-ft^2 floor area $\times$ 8-ft ceiling = 16000 ft^3 (450 m^3)

Forced-air gas furnace; no forced make-up air ($q_0 = 0$)

Gas stove:

    200 burner-min/day (Wade et al., 1975)

    75 oven-min/day (Wade et al., 1975)

    4 burner and 1 oven pilots, 24 hr/day

1 Smoker, 20 cigarettes smoked indoors/day

4 Adult occupants

1 Central electrostatic air cleaner, 3 airchanges/hr ($q_1 = 1350$ m^3/hr);

Chipboard walls, 1440 ft^2 total (133.8 m^2)

INFILTRATION

In this example, q_0 (forced ventilation make-up air rate) = 0. Infiltration is the only source of make-up air. Equation (6.6) is used to estimate q_2. Given wind velocities in the 5 to 20 mi/hr range, outdoor temperatures between 32 and 68°F, and the indoor temperature at 68°F, Equation (6.6) estimates q_2/V between 0.45 and 1.24 hr^{-1}. For this problem an average value of 0.75 is used and

$$q_2 = 0.75V$$

$$= 338 \text{ m}^3/\text{hr}$$

DESCRIPTION OF THE MODEL

The objective is to evaluate average air quality in a 7-room house. One approach would be to model each room of the house separately. This would require estimates of all parameters of the model [Equation (6.2)] for each room, a complex task. A review of the literature suggests this strategy is not necessary. Wade et al. (1975) made simultaneous measurements of NO, NO_2, and CO at four locations in a variety of homes equipped with gas stoves. Because of stove emissions, NO and NO_2 concentrations were higher in the kitchen then all other rooms. The maximum differences between rooms, however, were only a factor of 2. Carbon monoxide was approximately equal in all rooms. Palmes et al. (1977) made NO_2 measurements in the kitchen and a nonkitchen area of 19 homes. The average kitchen to nonkitchen difference in the homes with gas stoves was also only a factor of 2. Ju and Spengler (1981) studied the room-to-room variation in concentration of respirable particles in residences. While differences between some rooms were statistically significant ($p < 0.05$), the absolute differences were frequently very small and at most a factor of about 2. Episodic release of the tracer, sulfur hexafluoride, in the living rooms of 24 residences resulted in uniform indoor concentrations in adjacent rooms within 30–90 min (Moschandreas et al., 1978). Given only moderate room-to-room variations and the calculation of annual average concentrations, modeling each room separately is not justified.

The approach used is to evaluate the house as one large room (with a large surface area for pollutant decay and deposition) and assume a mixing factor, k, of 0.15. The annual average concentration is estimated by approximating steady state for all parameters and setting $t = \infty$. CO, CO_2, and NO concentrations are estimated using the pollutant mass balance model [Equation (6.3)] with sink $R = 0$. RP, NO_2, and O_3 concentrations are estimated using the pollutant mass balance model [Equation (6.4)] with a K term. K for particles is 0.05 hr^{-1} (Dockery and Spengler, 1981), 0.82 hr^{-1} for NO_2 (Wade et al., 1975), and 3.48 hr^{-1} for O_3 (Table 6.7). Derivation of O_3 and NO_2 decay rates are described in the following section *Decay and Deposition Factors*. Respirable particles are also estimated using the statistical model of Dockery and Spengler [Equation (6.20)]. Formaldehyde concentration is estimated using the pollutant mass balance model for outdoor air, smoking, gas-stove, and chipboard emissions [Equation (4.5)]. The value of $k = 0.15$ in Equation (4.5) is, and should be, the same as that used in the mass balance model [Equation (6.3)].

The central electrostatic air cleaner has a 99% efficiency (F_1) for tobacco smoke particles. Flow through the cleaner (q_1) is 800 cfm. F_0 and F_1 for all other pollutants are zero. Table 11.1 is a summary of parameter values for the model.

SOURCES

Table 11.2 shows sources, pollutants of concern, and emission rates. Outdoor air concentrations are average values for Chicago. Emission factors are from Chapter 4.

RESULTS

Table 11.3 shows the predicted concentration using the pollutant mass balance and statistical model. Predicted concentrations of CO_2 (0.33%), CO (8.1 ppm), and NO_2 (85 $\mu g/m^3$) are equal to or slightly below recommended standards (Tables 1.2 and ASHRAE, 1980). Predicted concentrations of RP (90 $\mu g/m^3$ with air cleaner) and NO (1.2 mg/m^3) are above recommended levels (Tables 1.2 and 7.1). The HCHO concentrations exceed the guidelines recommended in Table 7.1. Operating the electrostatic air cleaner increases O_3 levels by a factor of 5, but the O_3 concentration is very small. In addition, the air cleaner has a beneficial effect on RP concentration.

Table 11.1   Model Parameters for Evaluation of Air Quality Inside a Home

Parameter	Value
V	450 m^3
q_0	0
q_1	1350 m^3/hr
q_2	338 m^3/hr (see section on *Infiltration*)
q_3	q_2
q_4	0
F_0	0
F_1	99% for RP, zero for all other pollutants
K	zero for CO, CO_2, NO, and HCHO; 0.05 hr^{-1} for RP; 0.82 hr^{-1} for NO_2; 3.48 hr^{-1} for O_3
k	0.15

Table 11.2 Emission Inventory and Sources for a Residence

Source	RP ($\mu g/m^3$)	CO (mg/m^3)	CO_2 (g/m^3)	NO ($\mu g/m^3$)	NO_2 ($\mu g/m^3$)	HCHO ($\mu g/m^3$)	O_3 ($\mu g/m^3$)
Outdoor Air	$50^{a,b}$	2.9^a	1.3^c	125^d	63^a	7^e	29^a
	(mg/day)	(g/day)	(g/day)	(mg/day)	(mg/day)	(mg/day)	(mg/day)
Gas stove[f]							
Range top (8350 kcal/day)[g]	14.2	4.18	1670	835	501	59	
Range pilots (3840 kcal/day)[h]	—	0.48	750	154	96	—	
Oven (2500 kcal/day)[i]	—	1.25	500	225	125	28	
Oven pilot (960 kcal/day)[j]	—	0.96	190	2	48	—	
Smoking (one smoker, 20 cig/day)[k]	516.0	0.75	6.40	46	12.5	26	
Electrostatic air cleaner[l]							150°
People (4 people, 75% indoor)[m]			2520				
Chipboard walls (134 m²)[n]						293	

[a]Table 3.1.
[b]Wadden et al, 1980
[c]Chapter 3. *Outdoor Contributions*
[d]NAS (1977).
[e]Table 3.3
[f]Table 4.1
[g](2500 kcal/burner · hr) (3.33 burner · hr/day).
[h](40 kcal/pilot · hr) (4 pilots) (24 hr/day).
[i](2000 kcal/hr) (1.25 hr/day).
[j](40 kcal/hr) (24 hr/day).
[k]Table 4.5: (1 smoker, 20 cig/day).
[l]Table 4.12: 24 hr/day operation.
[m]Tables 4.16 and 7.3 (4 people, 1 met, 75% indoor).
[n]Equation (4.5): Chipboard on inside of outer walls only. House dimensions are 40 ft × 50 ft × 8 ft. This results in an α of 0.3 m⁻¹. $T = 18°C$,
$H = 27.4$ g/kg dry air, $n = 0.75$ hr⁻¹.
[o]Average of 8 values from Allen et al. (1978) (Table 4.12).

203

Table 11.3   Summary of Indoor-Outdoor Steady State Air Concentrations

Source	RP ($\mu g/m^3$)	CO (mg/m^3)	CO_2 (g/m^3)	NO ($\mu g/m^3$)	NO_2 ($\mu g/m^3$)	HCHO ($\mu g/m^3$)	O_3 ($\mu g/m^3$)
Outdoor air	50	2.9	1.3	125	63	7	29
Indoor air							
Pollutant mass balance							
With air cleaner	90.0	9.2	5.9	1200	85	340	4.8
Without air cleaner	337	9.2	5.9	1200	85	340	0.9
Empirical model							
Equation (6.20)	67.6[a]	–	–	–	–	–	–

[a] Air conditioning variable $A = 0$; smoking and outdoor contributions only.

For comparable conditions (no electrostatic air cleaner or air conditioning), the pollutant mass balance model predicts an RP concentration 5 times greater than the empirical model [Equation (6.20)]. Inspection of the steady-state mass balance [Equation (6.5)] provides an explanation of this difference. For large values of S, $C_{i,ss}$ will be most dependent on the values of k and q_2 in the denominator. Hence, a factor-of-2 change in k will change $C_{i,ss}$ by about the same amount, and similarly for q_2. No explicit information on values of these variables was provided in the Harvard study (Dockery and Spengler, 1981), but a variation by a factor of 2 in our assumed values of k and q_2 is not unreasonable. This exercise points up again the necessity for comparing estimated concentrations with actual measurement whenever possible.

DECAY AND DEPOSITION FACTORS

The removal term R [Equation (6.2)] can be evaluated as a deposition constant or a reaction rate constant. The form of R in terms of a decay rate is

$$R = KVC_i \qquad (10.4)$$

where the dimensions of K are time^{-1}. In this form, R is a function of K and the amount of mass present (VC_i). The form of R for deposition is

$$R = K_{dep}AC_i \qquad (6.12)$$

where K_{dep} is the deposition velocity and A is the area of contact. In this form, R is a function of K_{dep}, the area of contact and the pollutant concentration. The deposition velocity has the dimensions of length/time. The relationship between K and K_{dep} is then

$$K_{dep} = K \frac{V}{A} \qquad (6.13)$$

If removal is constant (not a function of C_i) then the solution of Equation (6.3) can be used. If removal is a function of concentration, Equation (6.4) is appropriate. In the following discussion $E = KV$ in Equation (6.4).

The decay factor for NO_2, K_{NO_2}, is derived from the data on the half-lives of CO and NO_2 simultaneously measured in a house where $q_0 = q_1 = 0$ (Wade et al., 1975). CO is a conservative pollutant and is dissipated from the house only by infiltration and exfiltration. NO_2 is a nonconservative pollutant and is simultaneously dissipated from and decayed inside the house. A model to describe this system is

$$V \frac{dC_i}{dt} = -kqC_i - KVC_i \tag{11.1}$$

where q = infiltration and exfiltration rate, volume/time
$\quad k$ = incomplete mixing factor
$\quad K$ = pollutant decay or deposition rate, time^{-1}.

Solving Equation (11.1) for NO_2, we obtain

$$\frac{C_i}{C_s} = e^{-(kq + K_{NO_2}V)\,t/V} \tag{11.2}$$

Solving Equation (11.1) for CO ($K = 0$), we obtain

$$\frac{C_i}{C_s} = e^{-kqt/V} \tag{11.3}$$

The half-lives ($C_i/C_s = 0.5$ at t = half-life) for measured values of CO and NO_2 were 2.1 and 0.6 hr, respectively (Wade et al., 1975). Substituting into Equations (11.2) and (11.3) and solving the decay factor for NO_2, $K_{NO_2} = 0.82$ hr^{-1} (0.014 min^{-1}).

The deposition velocity for NO_2 can be calculated from Equation (6.13). Using the data of Mueller et al. (1973) for the area to volume ratio in a home ($A/V = 0.8$ ft^{-1} or 0.026 cm^{-1}), the NO_2 deposition velocity is

$$K_{dep,NO_2} = (0.014 \text{ min}^{-1})(38.46 \text{ cm}) \tag{6.13}$$

$$= 0.54 \text{ cm/min}$$

The ozone decay rate K_{O_3} is based on the ozone deposition velocity for an office $K_{dep,O_3} = 2.23$ cm/min (Table 6.7) times the area-to-volume ratio of (0.8 ft^{-1} or 0.026 cm^{-1}) (Mueller et al., 1973).

For ozone, the decay factor is

$$K_{O_3} = K_{dep,O_3} \left(\frac{cm}{min}\right) \frac{A}{V} (cm^{-1}) \tag{6.13}$$

$$= 0.058 \text{ min}^{-1} \quad \text{or} \quad 3.48 \text{ hr}^{-1}$$

REFERENCES

Allen, R. J., Wadden, R. A., and Ross, E. O. (1978). Characterization of potential indoor sources of ozone. *Am. Ind. Hyg. Assoc. J.* 39:466–471.

Anderson, I., Lundquist, G. R., and Molhave, L. (1975). Indoor air pollution due to chipboard used as a construction material. *Atmos. Environ.* 9:1121–1127.

ASHRAE (1980). *ASHRAE 62-73R: Standards for Ventilation Required for Minimum Acceptable Indoor Air Quality.* The American Society of Heating, Refrigerating, and Air Conditioning Engineers. New York.

Dockery, D. W. and Spengler, J. D. (1981). Indoor–outdoor relationships of respirable sulfates and particles. *Atmos. Environ.* 15:335–343.

Ju, C. and Spengler, J. D. (1981). Room-to-room variations in concentration of respirable particles in residences. *Environ. Sci. and Technol.* 15:592–596.

Moschandreas, D. J., Stark, J. W. C., McFadden, J. E., and Morse, S. S. (1978). *Indoor Air Pollution in the Residential Environment,* vols. I and II, U.S. Environmental Protection Agency Report No. EPA 600/7-78-229a and b.

Mueller, F. X., Loeb, L., and Mapes, W. H. (1973). Decomposition rates of ozone in living areas. *Environ. Sci. Technol.* 7:342–346.

NAS (1977). *Medical and Biological Effects of Environmental Pollutants: Nitrogen oxides.* National Academy of Sciences, Washington, D.C.

Palmes, E. D., Tomczyk, C., and DiMattio, J. (1977). Average NO_2 concentrations in dwellings with gas or electric stoves. *Atmos. Environ.* 11:869–872.

Wadden, R. A., Allen, R. J., Scheff, P. A., and Hogan, T. J. (1980). Aerosol size characteristics in Chicago air, in press. *Proc. 5th International Clean Air Conference.* Buenos Aires, Argentina.

Wade, W. A., Cote, W. A., and Yocom, J. E. (1975). A study of indoor air quality. *J. Air Pollut. Control Assoc.* 25:933–939.

INDEX